# Inclusive Design Guidelines for HCI

# Inclusive Design Guidelines for HCI

Edited by
Colette Nicolle and Julio Abascal

London and New York

First published 2001 by Taylor & Francis
11 New Fetter Lane, London EC4P 4EE

Simultaneously published in the USA and Canada
by Taylor & Francis Inc.
29 West 35th Street, New York, NY 10001

*Taylor & Francis is an imprint of the Taylor & Francis Group*



Printed and bound in Great Britain by
Biddles Ltd, Guildford and King's Lynn



**Publisher's note**
This book has been prepared from camera-ready copy provided by the authors.

*British Library Cataloguing in Publication Data*
A catalogue record for this book is available from the British Library

*Library of Congress Cataloging in Publication Data*
Nicolle, Colette.
Inclusive design guidelines for HCI / Colette Nicolle & Julio Abascal.
p.cm.
Includes bibliographical references and index
1. Human-computer interaction. 2. Computers and the handicapped.
I. Abascal, Julio. II. Title.

QA76.9.H85 N53 2001
004'.01'9–dc21 2001023421

ISBN 0-7484-0948-3

# Contents

## PART 4: EXISTING GUIDELINES



## PART 5: GUIDELINES FOR SPECIFIC APPLICATION AREAS



## PART 6: THE FUTURE

# Authors

Prof. Julio Abascal
Informatika Fakultatea (School of Informatics)
Euskal Herriko Unibertsitatea (University of the Basque Country)
Donostia–San Sebastián, Basque Country
Spain
Tel: +34 94 301 8067
Fax: +34 94 321 9306
julio@si.ehu.es

Dr Demosthenes Akoumianakis
Foundation for Research and Technology - Hellas (FORTH)
Institute of Computer Science (ICS)
Human-Computer Interaction and Assistive Technology Laboratory
Science and Technology Park of Crete
Heraklion, Crete
GR - 71110 Greece
Tel: +30 81 391 743
Fax: +30 81 391 740
demosthe@ics.forth.gr
www.ics.forth.gr/proj/at-hci/cs.html

Dr Gary Burnett
Lecturer in Human Factors
School of Computer Science and Information Technology
University of Nottingham
Jubilee Campus
Wollaton Road
Nottingham NG8 1BB
United Kingdom
Tel: +44 (0) 115 951 3357
Fax: +44 (0) 115 951 4254
gary.burnett@cs.nott.ac.uk
www.cs.nott.ac.uk/~geb/

Andrew Downie
Teacher Consultant in Adaptive Technology
OTEN-DE Open Training Education Network—Distance Education
51 Wentworth Road, Strathfield
New South Wales
Australia 2135
Tel: 612 9715 8347
Fax: 612 9715 8345
andrew.downie@tafensw.edu.au
www.oten.edu.au/oten/

Prof. Jan Ekberg
Stakes (National Research and Development Centre for Welfare and Health)
P.O. Box 220
00531 Helsinki
Finland
Tel: +35 89 396 72091
Fax: +35 89 396 72054
jan.ekberg@stakes.fi
www.stakes.fi

Prof. Jan Engelen
Katholieke Universiteit Leuven
Dept. ESAT - Onderzoeksgroep Documentarchitecturen
Kasteelpark Arenberg 10
B-3001 Heverlee-Leuven
Belgium
Tel: +32 (0)16 32 11 23
Fax: +32 (0)16 32 19 86 or
+32 (0)16 23 74 31
Jan.Engelen@kuleuven.ac.be

Dr John Gill
Chief Scientist
Royal National Institute for the Blind
224 Great Portland Street
London W1N 6AA
United Kingdom
Tel: +44 (0) 20 7391 2371
Fax: +44 (0) 20 7388 7747
jgill@rnib.org.uk
www.tiresias.org

Luc Goffinet
Centre de Calcul
21 rue Grandgagnage
B-5000 Namur
Belgium
Tel: +32 81 72 50 38
Fax: +32 81 72 50 23
luc.goffinet@fundp.ac.be

Jan Gulliksen
Associate Professor
Department for Human Computer Interaction, Information Technology
Uppsala University
Lägerhyddsvägen 2, house 1
PO Box 337
SE-75105 Uppsala
Sweden
Tel: +46 (0) 18 471 2849
Fax: +46 (0) 18 471 7811
Jan.Gulliksen@hci.uu.se
www.hci.uu.se
also at
Center for User-Oriented IT-Design (CID)
Royal Institute of Technology (KTH)
Stockholm, Sweden
cid.nada.kth.se

Susan Harker
Senior Lecturer
Department of Human Sciences
Loughborough University
Loughborough
Leics. LE11 3TU
United Kingdom
Tel: +44 (0) 1509 263171
Fax: +44 (0) 1509 223940
s.d.harker@lboro.ac.uk
www.lboro.ac.uk/departments/hu/

Charles G. Hitchcock, Jr.
Chief Education Technology Officer and
Director, National Center on Accessing the General Curriculum
CAST, Inc.
39 Cross Street
Peabody
MA 01960
USA
Tel: +1 978 531 8555 x233
TTY: +1 978 531 3110
Fax: +1 978 531 0192
chitchcock@cast.org
www.cast.org/
www.cast.org/ncac/
www.cast.org/bobby/

Ann MacCann
Instructional Designer
OTEN-DE Open Training Education Network—Distance Education
51 Wentworth Road
Strathfield
New South Wales
Australia 2135
Tel: 612 9715 8192
Fax: 612 9715 8162
ann.maccann@tafensw.edu.au
www.oten.edu.au/oten/

Prof. dr. ir. Floris van Nes
IPO, Center for User–System Interaction
Technische Universiteit Eindhoven (TUE)
Den Dolech 2
P.O. Box 513
5600 MB Eindhoven
The Netherlands
Tel: +31 40 247 52 33
Fax: +31 40 243 19 30
f.l.v.nes@tue.nl
www.ipo.tue.nl/ipo/

Colette Nicolle
Research Fellow
HUSAT Research Institute
Loughborough University
The Elms, Elms Grove
Loughborough
Leics. LE11 1RG
United Kingdom
Tel: +44 (0) 1509 611088
Fax: +44 (0) 1509 234651
c.a.nicolle@lboro.ac.uk
www.lboro.ac.uk/research/husat/index.html

Prof. Monique Noirhomme-Fraiture
Institut d'Informatique
21 rue Grandgagnage
B-5000 Namur
Belgium
Tel: +32 81 72 49 79
Fax: +32 81 72 49 67
mno@info.fundp.ac.be
www.info.fundp.ac.be/~mno/

Knut Nordby
Senior Research Scientist
External Relations
Telenor Research and Development
P.O. Box 83
N-2027 Kjeller
Norway
Tel: +47 63 84 86 30
Fax: +47 63 81 00 76
knut.nordby@telenor.com

Prof. Helen Petrie
Director
Sensory Disabilities Research Unit
Psychology Department
University of Hertfordshire
Hatfield, Herts. AL10 9AB
United Kingdom
Tel: +44 (0) 1707 284 629
Fax: +44 (0) 1707 285 059
h.l.petrie@herts.ac.uk
www.psy.herts.ac.uk/sdru/hmpage.html

Dr David F. Poulson
Research Fellow
HUSAT Research Institute
Loughborough University
The Elms, Elms Grove
Loughborough
Leics. LE11 1RG
United Kingdom
Tel: +44 (0) 1509 611088
Fax: +44 (0) 1509 234651
d.f.poulson@lboro.ac.uk
www.lboro.ac.uk/research/husat/index.html

Patrick Roe
Lausanne Federal Institute of Technology (EPFL),
Laboratory of Electromagnetism and Acoustics (LEMA)
CH-1015 Lausanne
Switzerland
Tel: +41 21 693 46 31
Fax: +41 21 693 26 73
patrick.roe@epfl.ch

John W. Steger
Program Manager, IBM Special Needs Programs
jsteger@us.ibm.com

Prof. Constantine Stephanidis
Foundation for Research and Technology - Hellas (FORTH)
Institute of Computer Science (ICS)
Human-Computer Interaction and Assistive Technology Laboratory
Science and Technology Park of Crete
Heraklion, Crete
GR - 71110 Greece
Tel: +30 81 391 741
Fax: +30 81 391 740
cs@ics.forth.gr
www.ics.forth.gr/proj/at-hci/cs.html

Clas Thorén
Clas Thorén Consulting
Knalleborgsvagen 8E
SE-178 35 Ekero
Sweden
Tel and Fax: +46 8 560 355 05
clas.thoren@ctconsult.se

Dr Gregg Vanderheiden
Professor, Human Factors
Department of Industrial and Biomedical Engineering
Director, Trace R&D Centre
University of Wisconsin
S-151 Waisman Centre
1500 Highland Avenue
Madison WI 53705-2280
USA
Tel: +1 608 262 6966
Fax: +1 608 262 8848
gv@trace.wisc.edu
http://trace.wisc.edu/

Dr Carlos A. Velasco
Senior Researcher, GMD—German National Research Center for Information Technology
Institute for Applied Information Technology (FIT.HEB)
Research Division on Human Enabling
Schloss Birlinghoven
D-53757 Sankt Augustin
Germany
Tel: +49 2241 142609
Fax: +49 2241 142146
Carlos.Velasco-Nunez@gmd.de or velasco@pobox.com
http://access.gmd.de/

Tony Verelst
Chairman, ISdAC International Association
c/o IGL
Boddenveldweg 11
3520 Zonhoven
Belgium
Tel: +32 11 819465
Fax: +32 11 819466
Direct (video) phone: +32 11 823705 (Mon-Thu 12.30-16.00)
chairman@isdac.org
www.isdac.org/

Dr Neil Waddell
Research Fellow
HUSAT Research Institute
Loughborough University
The Elms, Elms Grove
Loughborough
Leics. LE11 1RG
United Kingdom
Tel: +44 (0) 1509 611088
Fax: +44 (0) 1509 234651
f.n.waddell@lboro.ac.uk
lboro.ac.uk/research/husat/index.html

# Foreword

Brian Shackel
Founder Director of HUSAT Research Institute
and former Chairman of IFIP TC 13

Imagine yourself, dear reader, in Shakespeare's seventh age "sans teeth, sans eyes, sans taste, sans everything" (*As You Like It*, II, vii, 165)—but of course you do not wish to live with any such imagining. Nor do I at the age of 74. Well, imagine coming in to work tomorrow, or going to your home television, and suddenly you simply cannot operate the remote TV control or you cannot even switch on your office computer and word processor; why—because your hands are all bent double with arthritis like that person you saw on the bus last week, or on that TV programme, trying with great difficulty to pay the bus fare.

Of course you can hardly imagine it; while we are fit and well, we neither wish nor are able to envisage what old age or any sort of disablement may bring. Moreover, the many facilities and benefits provided by advances in technology over the past 50 years tend to make us assume that there will be solutions ready and waiting if ever we become old or disabled. That assumption is well supported by examples—at least in the technological and functional sense.

But many products of these new technologies are designed by and can fairly easily be used by the healthy, young technologists who created them. The problem is that these brilliant technologists cannot imagine themselves into the physical and mental shortcomings that age or disability may bring. However hard they try, such imaginative leaps are very difficult. Therefore, the products are very often not at all easy to use.

So what to do? Recognise that designing for the elderly and disabled is a difficult and challenging intellectual process. If you fail, you have not failed imaginatively or emotionally; you have quite simply failed as a designer. You have failed to analyse, understand and allow for the constraints and inadequacies which beset the target users of your product. You have failed to search and find the relevant knowledge now available from ergonomics and human–computer interaction. You have failed to read and use the relevant advice in this book.

Here you will be directed towards a wealth of guidelines, developed from substantial research funded by the European Community, by US funding agencies, and by various national sources. Here is the advice which can help you to design successfully for elderly and disabled users. But here also you will learn that you have to put out some real effort—there are tools, but basically YOU must learn how to use guidelines effectively. Thus you will learn and be able to produce products and systems which may not, in fact, be quite 'As You Like It', but far more important they will be As The Users Like It.

# Acknowledgements

*Inclusive Design Guidelines for HCI* is the result of a long process and many interactions with the experts who kindly agreed to contribute. We wish to thank the authors for their enthusiasm, as well as their patience in meeting our various requests for more material or detail.

We are also grateful to:

- Professor Emeritus Brian Shackel, former IFIP TC 13 Chairman, for his Preface and also for his support from the book's early stages
- Our colleagues and friends of IFIP TC 13 and IFIP WG 13.3 for their help, in particular to the current TC 13 Chairperson, Judy Hammond
- Tony Moore, Senior Editor at Taylor & Francis in London, for his never-ending guidance
- Leela Damodaran, Head of the Human Sciences and Advanced Technology (HUSAT) Research Institute at Loughborough University, for her encouragement and support
- Kathy Phillips, Information Scientist at HUSAT, for her invaluable contributions with the formatting and proofreading of the manuscripts, and
- Colette's husband David and Julio's wife Jerusalem for their understanding when the time we spent on the book should have been with them and the family!

# Part 1

# Introduction

CHAPTER ONE

# Why Inclusive Design Guidelines?

Julio Abascal and Colette Nicolle

## 1.1 INTRODUCTION

Let us say you wish to design and develop a new product or technology. You want to ensure that as many people as possible are able to use it—not only is it politically and in some countries legally correct but with our ageing population it makes good economic sense. You want to follow existing design advice, of which you have been told there is much around. You wish to follow an *inclusive* or *universal design* philosophy, creating a product or system which will be usable by as many people as possible, including people who are older or disabled.

Where do you look for design advice? You may decide to look for published materials, such as design guidelines. However, when you find guidelines, can you be sure that they will be applicable to your specific product or technology? Do existing guidelines take into consideration the needs of all possible users? If you follow these guidelines, can you be sure that your product will be usable by more people?

As Gregg Vanderheiden describes it, *universal design* is the process of creating products (devices, environments, systems and processes) which are usable by people with the widest possible range of abilities, operating within the widest possible range of situations (environments, conditions and circumstances).[1] To try to make a product usable by *everyone* is a near impossible task—but we can try to exclude as few people as possible by ensuring that products and systems are flexible enough to be adapted to individual needs and preferences. Including as many people as possible also defines for us the term *inclusive design*—a *design for all* or *universal design* which does not exclude them from using the system. In other words, this means products without added barriers.

Older and disabled people may be the most likely to benefit from new technologies, products and services, but in reality may have difficulties in taking advantage of the systems due to the very same limitations. Thus, as advances are made in areas such as information and communication technology (ICT), older and disabled people may lag behind—unless technologies are designed with their functional impairments and their requirements in mind. So, just as everybody loves a 'curb-cut'[2] (or slope in the curb), like the parent with a stroller or the child on a skateboard, it is not just the older or disabled person who will benefit from better guidelines. Each one of us can become handicapped in certain environments, for example using controls when our hands are cold or when the lighting is low, or

---

[1] See trace.wisc.edu/docs/whats_ud/whats_ud.htm

[2] Term used by Gregg Vanderheiden, 13 October 1983, Keynote Address, Computers: The Greatest Single Handicapping Condition of the Future? *Annual Governor's Conference on the Handicapped,* Indianapolis, Indiana, USA.

more generally when travelling in a foreign country. So designing products that are usable by older and disabled people will also help to ensure that it is easier and more convenient for everyone to use them.

A difficulty with guidelines, however, is that by their very nature they are simplifications that must be general enough to be applicable to a wide range of products and technologies. They are usually drawn from best available practice, sometimes applied to different situations and technologies and often not validated for your own specific area. Where guidelines are more precise, for example by specifying particular dimensions, they are difficult to apply to other technical areas or are considered too restrictive by innovative designers. Where technological innovation is taking place, there may be no guidelines available and a designer may end up drawing from inappropriate design advice without questioning its validity. Frequently, it is also true that design recommendations are conflicting, not only between different sets but also within the same set of guidelines.

Design guidelines have been frequently used in different environments for diverse purposes such as the storage of knowledge and the transmission of successful experiences among designers. Such experiences prove that sets of guidelines also help to maintain coherence in large design teams. For this reason, many teams are in the habit of recording their design-validated decisions to use them in future systems. Many types of guidelines exist. This book is centred on the ones devoted to the design of systems that will be accessible to everybody, including older and disabled users.

## 1.2 BACKGROUND TO THE BOOK

In 1992 the International Federation for Information Processing (IFIP)[3] accepted the proposal issued by Technical Committee 13, devoted to human–computer interaction (HCI), to create a new working group dedicated to 'Human–Computer Interaction and Disability' (for which the editors of this book were Secretary and founder Chairman). WG 13.3 had its first meeting on 20th–21st November 1993 in Donostia-San Sebastián, where the aims and scope were agreed. The goals of WG 13.3 include the following:

- To make HCI designers aware of the needs of people with disabilities
- To monitor the latest developments in the design of HCI and their impact on accessibility and usability
- To recommend guidelines for the design of HCI to facilitate the use of computers by people with disabilities.

In the WG 13.3 meeting held on 14th October 1995 in Lisbon, our interest in design guidelines was discussed. Since there is much design advice already available, we decided to study, promote and disseminate knowledge about design guidelines for accessible systems through various activities, such as the organisation of a series of workshops[4] on this topic.

---

[3] www.ifip.or.at/

[4] Workshop *WG13.3 HCI and Disability*, held 26th June 1995 at INTERACT '95 (Lillehammer, Norway); Workshop *Guidelines for the Design of HCI Systems for People with Disability*, held 15th July

The Workshop during INTERACT '97 in Sydney collected and discussed previously issued sets of guidelines, along with the best way to classify and disseminate them. The idea for this book stemmed from the presentations at that workshop, although most chapters have been written specifically for this publication. The purpose was to compile some interesting experiences to help designers to find, select, understand and use design guidelines, as well as their related techniques and tools.

## 1.3 USING GUIDELINES

Using guidelines is expected to make the design process easier and to maintain coherence with previously taken design decisions, but using guidelines is fairly difficult. The stored knowledge is not always clear and it requires some interpretation. Guidelines can be ambiguous, contradictory or only partially true. The person using guidelines must have a clear idea about the design objectives and must have the capacity to decide whether a concrete guideline can be applied or not.

Thus, the designer who uses guidelines to check the usability of his or her design faces problems such as:

- *Finding sets of guidelines that are relevant and adequate to the actual design:* It is not difficult to find guidelines published in books and journals. However, Web search is most convenient because more recently updated guidelines can be found there.
- *Choosing the set of guidelines that best fits the characteristics of the design:* When more than one set of guidelines are found, the design team must select one or more among them to be used for the design process. This selection should be done in relation to the ability of the sets to fit the aims and the environment of the intended design. If more than one set of guidelines are used, it is necessary to have checked their compatibility.
- *Checking the soundness and reliability of the guidelines:* It is also important to know if these guidelines have been verified with real users or, as happens in many cases, if they are just intuitive ideas.
- *Checking the coherence and applicability of the guidelines:* Once we have selected one or more sets of guidelines, they may contain contradictions and incoherence. How to solve these contradictions when they appear must be clarified beforehand.
- *Finding an adequate design methodology that is compatible with the use of guidelines during different stages of the design process:* Among the diverse design methodologies, the ones that can include recommendations and guidelines in the design will be able to facilitate conceptualisation and the achievement of objectives.

---

1997 at INTERACT '97 (Sydney, Australia); and Workshop *Making Designers Aware of Existing Guidelines for Accessibility*, held 31st August 1997 at INTERACT '99 (Edinburgh, United Kingdom).

- *Coping with a large number of guidelines:* As we have mentioned before, sets of guidelines are frequently compiled from the experience of large working groups in diverse designs and tend to incorporate different experiences coming from these designs. The sets of guidelines may become very large,[5] and the know-how they contain does not necessarily increase with the number of guidelines stored within them. A modular approach to the use of guidelines, separating the different contexts, groups of users, goals, etc., very much simplifies their application.
- *Using convenient design tools that facilitate intensive implementation work:* It is highly recommended to select a methodology compatible with guidelines that facilitate their application to the design process.

Each of these issues is covered by one or more chapters of this volume, from diverse points of view and at different levels, trying to help the interested reader adopt a successful approach to the use of inclusive design guidelines. For instance, Stephanidis and Akoumianakis provide a comprehensive discussion of these issues and Poulson gives some interesting examples of such inconsistencies from technologies in the home automation sector. Gill also provides examples where recommendations are no longer valid when the technology changes, e.g. from analogue to digital television.

## 1.4 TRAINING AND TOOLS

Training in the use of guidelines is crucial. Frequently, however, guidelines have been considered as rules of thumb—the opposite of formalisation—and therefore have been ignored by academics. Thus, junior engineers working in a design team have to cope with guidelines with almost no previous experience—experience, we feel, being the best way to learn the use of guidelines. In order to avoid long training times, however, teaching methods need to be developed which will shorten the time required by 'natural' learning. One useful technique to teach students to use guidelines is to put them in a 'pretend' design context and encourage them to take reasoned design decisions, based on the knowledge they can extract from guidelines. This book also aims to help in the process of learning the use of guidelines, offering good usage examples, showing existing sets of guidelines and also through the discussion of various methods and tools.

Throughout the book, the reader will realise that working with guidelines is not an easy matter. To be able to face some of the previously mentioned difficulties, training is not enough. The huge number of guidelines and the difficulty of their application points to the need for sound and powerful methodologies that simplify the design process. Likewise, there is clearly a need for design tools with integrated facilities to handle guidelines. We hope that this book will promote such development—and in turn, an effective strategy for applying guidelines in the design process.

---

[5] As Nielsen (1993) says, "Current collections of usability guidelines typically have on the order of a thousand rules to follow and are therefore seen as intimidating by developers."

Valuable efforts have been made everywhere to study and discuss this issue. For instance, the International Special Interest Group on Tools for Working with Guidelines has organised successive workshops on this topic—the last one was held in Biarritz in October 2000 (Vanderdonckt and Farenc, 2000). From the conviction that inclusive design guidelines can be adequately applied only if good methods and tools are available, three chapters in this book refer to tools: Chapter 6 written by Stephanidis and Akoumianakis, Chapter 7 from Goffinet and Noirhomme-Fraiture and Chapter 8 from Hitchcock. In addition, Chapter 11 by Poulson and Waddell introduces USER*fit*, a methodology for user-centred design mainly focused on assistive technologies.

## 1.5 USER PARTICIPATION

Most HCI experts encourage user participation in the design and test processes. This participation is often reduced or even omitted because it is expensive in terms of time, human resources and finances, and because the technical staff and the users may have difficulties in understanding each other's views.

In working towards inclusive design, the participation of the user is even more crucial because frequently designers overlook the special needs they should be taking into consideration.[6] It is important to avoid supposed and intuitive knowledge about the user, often coming from the extrapolation of experience, beliefs and clichés in the minds of the designers.

One of these clichés is technology aversion among disabled and elderly people. Many designers think that elderly people reject technology outright. Nevertheless, some studies show that adequately trained elderly people are in general able to use technology.[7] In addition, it is expected that the acceptance of technology by elderly people will increase with the ageing of people accustomed to technology.

We cannot forget that technology rejection, when it exists, is most likely caused by poorly designed interfaces or by insufficient evaluation of user needs. Poor quality interfaces can lead to difficult, stressful and even unsafe use of a device or service. Furthermore, ignorance about user needs encourages engineers to imagine hypothetical benefits for disabled and elderly people when they are applying emerging technological advances to solve presumed user needs. These assumptions, when they are not founded on users' interests, needs, wishes, likes, etc., lead to systems which are not suited to the target user group and hence are likely to be rejected. Involving users will result in systems that satisfy the true user needs and are more likely to be accepted by them.

Therefore, there are many reasons why inclusive design guidelines should include input from real users. Chapter 2, by Velasco and Verelst, contains interesting thoughts and experiences about user participation as part of the Information

---

[6] For details about user participation in assistive technology research, see the results of the European project FORTUNE at www.fortune-net.org

[7] *Attitudes and Acceptance* by S. Bjørneby *et al.* (1991) refutes the myth of the technological incompetence and disinterest of elderly and disabled people to use technology from experiments carried out in England, Portugal and Norway.

Society disAbilities Challenge International Association (ISdAC) and within other European and national projects.

## 1.6 THE ROLE OF LEGISLATION

There is now clearly an increased awareness of the rights of people with disabilities, leading to an increased importance of design guidelines that promote equal access. The USA has made great progress in this area with the Americans with Disabilities Act or ADA.[8] In the European context the rights of all citizens to equal opportunities have also been stressed in the European Commissions (1996) White Paper—*Equal Opportunities and Non-Discrimination for People with Disabilities.*

In 1995 in the UK, the Disability Discrimination Act or DDA[9] introduced new laws and measures aimed at ending the discrimination which many people with disabilities face, giving them new rights in the areas of employment, goods, facilities, services, property, transport and education. For example, the DDA says that reasonable adjustments may have to be made to employment arrangements or premises, e.g. by acquiring or making changes to equipment. This might include providing appropriate assistive devices which would enable a person to use a computer. More importantly, however, it may point towards a need for a design philosophy which will ensure that systems meet the needs of older people and people with disabilities, as far as possible, from the outset.

To apply current legislation, the designer needs to interpret it in terms of technical development. This task is not always an easy one because the lawmaker and the technician usually do not speak the same language. Guidelines can help to obtain a design compliant with legislation. Moreover, to facilitate this process, some laws can take the form of technical guidelines.

Frequently standards also have the form of very precise guidelines—often because they initially came from widely accepted guidelines. However, as Nordby so clearly describes in Chapter 5, legislation may well be the only way to encourage industry to comply with guidelines and standards for inclusive design.

## 1.7 SOCIAL AND ETHICAL ISSUES

Social and ethical issues are usually expected to be outside the scope of the technical designer. His or her task is to translate accurate, detailed and unambiguous specifications into working devices or services. These specifications are expected to be in line with ethical and social aspects, but frequently they are ambiguous or even inaccurate and therefore accept different implementations, some of which can be socially intrusive or ethically inadmissible (Abascal, 1997). To be able to take certain design decisions, the designer should also have advice about these aspects.

For example, tagging is considered by some people as a suitable solution to reduce the risk to people with dementia who wander. The use of tagging technologies on the wrist or ankle of a person has raised many ethical, legal and cultural

---

[8] www.usdoj.gov/crt/ada/adahom1.htm
[9] www.disability.gov.uk/dda/index.html

issues, often exaggerated by the use of the term 'tagging' which gives the negative connotation of tracking offenders. People are very aware of individual rights and the danger of imposing regimes of care on those who are unable to give or refuse their consent. Nevertheless, tagging is considered by many carers to be the least unsatisfactory and the least objectionable alternative to protect the safety of the person with dementia. In Nicolle and Richardson (1995), recommendations were made for the design of tagging systems so that they would consider such ethical issues and the rights of the individual. For example, technology to be used in the care of people with dementia must be flexible enough to allow different levels of interface for different stages of dementia, as well as the ability to deactivate certain facilities, like an alarm button, for some people too confused to use them. The device also should not label the person, i.e. it needs to be light, discreet and aesthetically appealing. It is important to remember that a person with dementia should not be expected to wear a design which another person might reject.

So, even if it is not common to find guidelines on ethical and social considerations, we recommend that existing guidelines include advice to ensure that social inclusion, privacy and decision-making are not overlooked.

## 1.8 AIMS AND SCOPE OF THIS BOOK

This book illustrates the wide range of inclusive design guidelines that exist for human–computer interaction and where to look for further information. Other relevant terms may be used, however, including human–machine interaction and user–system interaction.[10]

The authors of this book introduce a range of existing sets of general guidelines, as well as guidelines specific to certain application areas. At the same time, they also investigate the issues discussed above and assess whether guidelines developed for specific users, activities and contexts of use can be relevant and applicable to other application areas.

We do not propose that this is an exhaustive study of existing guidelines for accessibility. In fact, we do not provide a detailed review of such guidelines as the Microsoft Windows® *Guidelines for Accessible Software Design*[11] and the Apple guidelines.[12] The reader can easily go to those Web sites for detailed information on implementing their accessibility features. Furthermore, new guidelines are constantly appearing, while others become obsolete due to advancements in technology. For this reason, the focus is oriented to more permanent issues about guidelines, such as their validity and availability, as well as applicable methodologies and tools to work with them effectively. Nevertheless, some existing sets of guidelines are presented and discussed to ensure better understanding of these concepts.

We believe that the guidelines and guidance presented here will contribute to an inclusive design philosophy leading to more usable systems for all. At best the book should direct the reader towards a more effective use of guidelines—or at

---

[10] In Chapter 5 on markets and regulations, Nordby stresses that 'computer' should be understood in the very widest sense, since complex consumer products will now very often have a built-in or 'embedded' microcomputer.

[11] www.microsoft.com/enable/dev/guidelines/software.htm

[12] www.apple.com/education/k12/disability/

least it will promote further discussion. We also expect that this book will contribute to the development of other related aspects, mentioned before, some of which are not frequently treated when studying the use of guidelines, for instance:

- the teaching of the use of guidelines for the design of accessible technology
- implementation of and adherence to accessibility legislation
- the need for awareness of ethical and social aspects, and
- the need for full user participation during all phases of the design cycle.

The book is divided into the following parts and a summary of each chapter is provided below:

Part 1: This Introduction
Part 2: General issues in the design process
Part 3: Tools for accessing and using guidelines
Part 4: Existing guidelines
Part 5: Guidelines for specific application areas
Part 6: The future

**Part 2: General issues in the design process**

In Chapter 2, Carlos Velasco and Tony Verelst deal with two processes that have a strong influence on the uptake of guidelines in the design of any device or service: training of designers and implementers of the guidelines, and verification and evaluation of the guidelines to confirm their validity and usefulness. This second stage is linked to the training of end-users, who can help to give useful feedback to designers. The chapter is based upon the experiences of the authors within the Information Society disAbilities Challenge International Association (ISdAC) and within other European and national projects.

In Chapter 3, Helen Petrie outlines some of the limitations of previous classifications of impairments, disabilities and handicaps as a source of information for the design of new technologies which will assist disabled and elderly people in their daily living. She proposes a classification based on an analysis of the functional capabilities of various human systems, including visual, auditory, motor and cognitive capabilities. This classification has been used to develop suggestions for the design of information and communication technologies (ICTs), which will make these technologies more accessible to disabled and elderly people, as well as often meaning a better design for everyone.

In Chapter 4, Floris van Nes deals with the validity of design guidelines for generic human–computer interaction and user–system interaction (USI), in particular for older people and users with disabilities. He stresses that USI design guidelines for older people or disabled users can only make a real impact by their being standardised—preferably internationally—followed by legislation based on these standards. Examples are given of such standards in telecommunication from ETSI and the ITU–T.

In Chapter 5, Knut Nordby shows how the market, competition, liberalisation, privatisation and deregulation of converging telecommunications and information technology—into information and communication technology or ICT—may result in poorer services and facilities for older and disabled people. He shows how guidelines, public procurement, regulations, standards and legislation can be used to protect users from any of these ill-effects. However, ultimately he sees no alternative to legislation in order to encourage industry to comply with guidelines and standards for inclusive design.

**Part 3: Tools for accessing and using guidelines**

In Chapter 6, Constantine Stephanidis and Demosthenes Akoumianakis present a method for managing accessibility guidelines during user interface design. In addition, a user interface design platform is presented which supports the methodology and allows designers to propagate accessibility guidelines into user interface implementation. Guideline management is considered in terms of three distinctive phases, namely consolidation, deposit and propagation. This design environment offers tools which support each one of these phases. Finally, the authors discuss example case studies and application experience in the context of their collaborative research and development efforts.

In Chapter 7, Luc Goffinet and Monique Noirhomme-Fraiture introduce the fact that most HCI guidelines are not really well presented on the World Wide Web (WWW), and when they can be found, they are generally available as only plain text without any common structure and without any cross-references. They present a way to retrieve relevant HCI guidelines on the Web using their specially configured meta-search engine. After gathering relevant guidelines and structuring them manually, they tackle the problem of automatically generating relevant cross-references between them, using statistical computations that are commonplace in the field of information retrieval. A human expert is also introduced to refine the process by assessing a sample of generated cross-references—relevance feedback is given to the computer so that better cross-references can be output by the system.

In Chapter 8, Charles Hitchcock tells us about the Bobby tool, emphasising that with a modest amount of effort, the WWW can be made accessible to everyone, including individuals with disabilities. This is best accomplished when Web sites are under development. For sites that have already been made available on the Internet or on local intranets, Bobby, developed by CAST in Peabody, Massachusetts, USA, will test pages for accessibility and guide the repair process. The chapter provides a general overview of Bobby, a description of the guidelines upon which it is based, technical notes, comments about how Bobby is being used, and future development plans.

### Part 4: Existing guidelines

In Chapter 9, Clas Thorén describes the *Nordic Guidelines for Computer Accessibility*, published by the Nordic Cooperation on Disability. This chapter provides a useful distinction between the procurement process and the design process and illustrates how the Nordic Guidelines are intended to be used in both contexts. The *Guidelines* nevertheless achieve an important balance between accessibility of personal computers and workstations for older and disabled people and the needs of procurers and suppliers for requirements that are easy to understand and evaluate.

In Chapter 10, Jan Engelen provides an historical account of the WWW and the international development of guidelines, through the Web Accessibility Initiative (WAI), to make it not just accessible but also usable for older and disabled people. He describes how such initiatives in the US and in several European countries have now led to the adoption of accessibility rights by the European Commission itself in the eEurope 2002 initiative. However, he stresses that many issues still remain unsolved, e.g., using the Internet on miniature hand-held computers, and that accessibility can be considered a never-ending story not different from the development of the Internet itself.

In Chapter 11, David Poulson and Neil Waddell describe the design methodology and manual called USER*fit*. This provides guidance on user involvement and user-centred design of assistive technology products, including methods to achieve it and a range of helpful recommendations and design advice. A key aspect of the methodology is that it forces design issues to be made explicit—it makes designers ask the right questions and justify and document their design assumptions or decisions, either about the technology or about its users. This strategy could be equally applied to the inclusive design of all types of products and services.

In Chapter 12, Jan Gulliksen, Susan Harker and John Steger report on the work currently being carried out by the International Organization for Standardization (ISO) on accessibility of human–computer interfaces. ISO Technical Specification 16071 *Ergonomics of human–system interaction—guidance on software accessibility* relates the concept of accessibility to usability. It provides guidance to complement traditional usability guidelines for the user interfaces to become usable to users with the widest possible range of capabilities. The chapter reports upon the process and results achieved during the production of this ISO working group.

### Part 5: Guidelines for specific application areas

In Chapter 13, Jan Ekberg and Patrick Roe highlight aspects of good design and accessibility for various telecommunication devices and services. They emphasise not only the importance of HCI guidelines to the telecommunications sector, but also the importance of accessibility to older people and people with disabilities. Work done by the European collaboration projects COST 219 and COST 219bis is described, including their development of general guidelines for accessible telecommunications products and services (e.g. with regard to reading a display), as well as guidelines covering specific application areas (e.g. mobile telephones). The

*Handbook on Inclusive Design*, developed by the EU INCLUDE project, gives designers concrete recommendations to follow the 'design-for-all' principle.

In Chapter 14, John Gill describes the variety of problems people who are older and disabled have with public access terminals, in particular cash dispensers. Recommendations are made not only for the design of the terminal's hardware and software, but also for the use of smart cards to store the user's particular requirements. This chapter clearly demonstrates that many such features which are essential for older and disabled people are advantageous for all users.

In Chapter 15, Colette Nicolle and Gary Burnett stress the importance of mobility for people who are older or disabled and how intelligent transport systems can make their travelling by private or public transport easier and more comfortable. As well as stressing the importance of the general HCI guidelines in this book, prescriptive and process guidelines are presented from the European Union's (EU) TELSCAN project. These include guidelines on how to design the system interface, what type of specific information is needed by older and disabled travellers, and what methods and tools can be used to ensure that the needs of older and disabled people are considered in the system evaluation process.

In Chapter 16, David Poulson provides an introduction to the range of products and services that are currently being explored in the home automation sector, with particular reference to supporting the needs of older and disabled users. He draws not only from general guidelines for interface design, but also on specific work from the EU TIDE CASA project. It is interesting to see that there has been little integration of the different systems in the home and that this is where home automation differs from existing home products. The performance of the whole home automation system, made up of a large number of elements, requires careful integration of each part.

In Chapter 17, Andrew Downie and Ann MacCann apply general design guidelines to the specific context of computer-based instruction and learning materials to ensure that people with disabilities can obtain maximum benefit. Although such advice is relevant to the development of even paper-based materials, special emphasis has been given to the specific features of electronic educational materials, drawing from the authors' experiences with students with physical disabilities and learning difficulties.

**Part 6: The future**

In the final chapter, Gregg Vanderheiden describes some current and future technologies that demonstrate the potential for new flexibility and built-in accessibility. However, he suggests that technologies in the future may be sufficiently different that specific techniques that we use to make products accessible today by retrofit or adaptation may not exist tomorrow. We, therefore, need to look towards general approaches for making products more flexible and easier to use for all. A 'first pass' summary table is provided of what these general principles and guidelines might look like.

# Part 2

# General Issues in the Design Process

CHAPTER TWO

# Training, Verification and Evaluation of Guidelines

Carlos A. Velasco and Tony Verelst

## 2.1 INTRODUCTION

We are in the middle of a revolution similar to that of the industrial revolution of the 19th century, the *Information Society*. This revolution is affecting the way we work, the way we study, the way we shop and many other aspects of our daily life. Concepts such as telework, e-commerce, cooperative work, mobile telephony, tele-cottage or tele-training are becoming part of our regular vocabulary. Furthermore, many appliances of our everyday life such as a dishwasher or a VCR will incorporate new functionalities aimed to make the life of the average user easier.

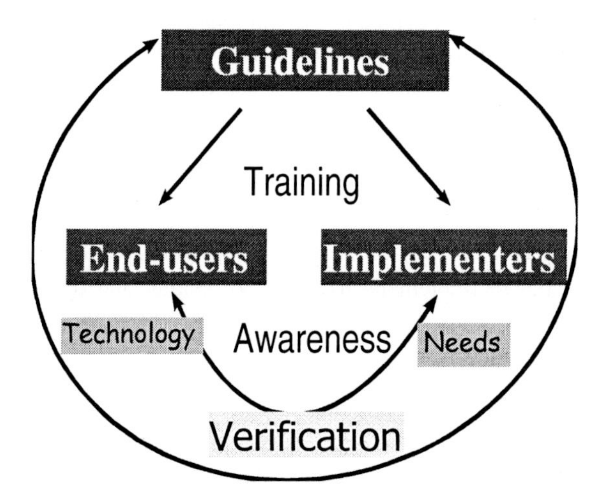


**Figure 2.1** Global overview of the verification process for design guidelines

This book is introducing the reader to a subset of guidelines which are intended to facilitate access to new or existing devices and services to a broader range of users, mainly disabled and older people. This chapter deals with two stages that have a strong influence on the uptake of these guidelines in the design process of any device or service: training of designers and implementers of the guidelines, and verification and evaluation of the guidelines in order to confirm their validity and usefulness. This second point is linked with the training of end-users, who can help to give useful feedback to designers. The chapter is based upon the experiences of the authors within the Information Society disAbilities Challenge International Association (ISdAC[1]) and within other European and national projects. ISdAC is a non-profit international association under Belgian law, aimed at promoting a full integration of people with disabilities in the information society by collaborating with international and national bodies and by spreading information about accessibility and assistive technology to its constituency. ISdAC membership is composed mainly of people with disabilities, with experience in the use and design of information society and assistive technology tools and services, together with a group of advisors with known experience in different sectors.

Figure 2.1 outlines our approach to this topic. This approach has been taken in several workshops[2] and we feel can lead to a dramatic improvement in the uptake of inclusive guidelines. The process can be divided into several steps. First, both end-users and implementers or designers are trained in different aspects of the guidelines (those relevant to them). After this initial phase, both groups undergo an 'awareness' phase, where end-users delve into some technical aspects whilst implementers become familiar with different issues related to assistive technology. Then begins a new phase of interchange between both groups that usually leads to new recommendations, i.e. a likely modification of some guideline, or to a standardised implementation process. We have divided the chapter into three main sections. The first section is devoted to the training phase of end-users. The second section will describe in detail the training phase for implementers. Finally, the third section will describe the verification/evaluation phase, with a practical example extracted from our experience.

## 2.2 TRAINING END-USERS

Whenever implementing guidelines, designers must evaluate whether the guidelines issued are amenable to implement and easy to handle, and the end-users of the product must:

- be satisfied with the performance of the interface resulting from the guidelines
- be aware of the technology limitations (state of the art) to be able to compare products
- be able to provide useful feedback to designers.

It is obvious that for the average end-user, it is not feasible to go beyond the first point. However, some authors (Brandt and Gjøderum, 1995) and our own experience have shown the benefits of promoting the three steps.

It is important to stress that a fruitful interaction and a constructive result can only be the product of common ground where end-users and designers can exchange opinions and experiences. As already mentioned, the education of the

[1] www.isdac.org/

[2] For example, it was used with success in an on-line workshop organised within the framework of the 3rd TIDE Congress held in Helsinki in June 1998.

average user is frequently outside the scope of technical fields. Thus there is not a wide test bed of end-users capable of providing useful feedback. Furthermore, we have found out from ISdAC that key international organisations of the sector lack the level of expertise one would expect.

The conclusion is that training programmes must be carried out in order to increase this test bed. Unfortunately, there are not too many programmes of this kind running at the moment. The FORTUNE project[3] is an excellent example to be imitated.

It must be stressed that the training has to be carefully designed. The major problem that can be found is a negative attitude from the end-user to the implementer or designer which arises from the fact that certain products place big hurdles towards access. The key point of any training programme is to be able to turn this attitude into useful criticism, and not into frustration.

Another issue, sometimes forgotten, is that the training of the end-user must not be focused on his/her own specific needs, but also on those of others, to facilitate a wider view to meet the demands of as many end-users as possible. Setting up a large test bed of users to rely on is a good approach concerning this matter.

In summary, the training of users must be focused on the following points:

- *Awareness about other types of disability*
- *Knowledge of the state of the art of existing technologies.* For example, if user agent guidelines[4] are to be evaluated for Internet navigation, the users must be aware of existing browsers and authoring standards[5]
- *Understanding of how guidelines improve the final device or service.* Following our example, users will compare the performance and usability of the browser with and without its implementation (probably by comparing with older versions of the same browser)
- *A search for additional elements not covered by the guidelines.* This point focuses on the search for additional features aimed at implementing a better functionality, or to deal with a lack of functionality
- *Understanding of the basics of the design process* to be able to interact with the implementer. In order to achieve this point within our example, users should be aware of the different operating systems and their various software development approaches.

## 2.3 TRAINING DESIGNERS AND IMPLEMENTERS

There are many references in the literature on guidelines to design almost any kind of product or device. However, when dealing with the training process on these guidelines, figures decrease sharply. The task of developing guidelines is difficult

---

[3] www.fernuni-hagen.de/FTB/fortune/welcome.htm

[4] www.w3.org/TR/UAAG/

[5] www.w3.org/TR/WAI-WEBCONTENT/

and time consuming but they will not succeed if their potential implementers are not trained appropriately.

Furthermore, guidelines are frequently focused on the end-user of the product and they do not consider within their approach the needs of the people who will implement them. That will certainly lead to the failure of the guidelines as well.

This section provides a set of recommendations resulting from our experiences in training software developers and Web designers, on different guidelines applicable to mainstream products to make them accessible to disabled and older people. We feel that these recommendations can easily be extended to designers and implementers of other products.

### 2.3.1 Is the concept *design for all* misleading?

The first challenge faced is the definition of 'Design for All' or 'Accessible Design'. There are many definitions—it is a widely discussed topic and we are sure that the audience is familiar with many of them (Poulson *et al.*, 1996; Janssen and van der Vegt, 1998; Sandhu, 1998). Concepts such as designing for the broader average can be vague enough to mislead the implementer of any guideline who is not familiar with assistive technology. Generally, it gives to the implementers the idea of a highly sophisticated and not cost-effective design process by which their products will reach a wider hypothetical market whose significance has not been shown to them.

In the field of Web design and software development, our experience shows that outreach and education benefit when the message is simplified. The message to be sent is that we can solve simple problems with simple solutions. Some of the problems and some of the solutions are not that simple, and the designer will be overwhelmed if he or she must tackle one hundred issues simultaneously. Instead, the focus should be on the concept of *universal design*, or 'good user-based design', which will address the needs of the broadest range of users by adding 'customisation' to the resulting interface (Story, 1998; Stephanidis *et al.*, 1999).

### 2.3.2 Awareness of the situation and real-life examples

The first step when training implementers on guidelines about accessibility issues is to make them aware of what issues and challenges a person with disabilities must face in dealing with computers or whatever product is under consideration. As pointed out in section 2.2, interaction with the end-user is the desirable situation but there is not always such an opportunity. Thus, alternatives to raise awareness must be sought. A typical example will be the use of computers by people with disabilities. We have found many times that software designers never thought of a person with disabilities using a computer. A simple exercise of awareness can help to implement guidelines from a different point of view.

It is helpful to show to the implementers real people interacting with computers in their normal environment. There are many video tapes[6] and other audio-visual material showing people with disabilities which can be quite helpful. These are available from different organisations and can be obtained at a minimum cost or even free of charge. This also can help the implementer to obtain a flavour of different assistive devices and what type of problems users face with them.

This drives us to other key questions within this awareness exercise. First of all, designers must be aware that inclusive design will increase their market potential. It is beyond the scope of this book to discuss the market for different products but, for example, in 1993 there was a potential market of 25 million disabled and elderly people in the European Union, growing to 32 million by the year 2020 (EUROSTAT, 1992; Carruthers *et al.,* 1993). The realisation of this potential is a shared task of the community and implementers should be aware of it. Furthermore, it must be emphasised that the use of inclusive guidelines benefits other users of your product, as will be pointed out in the following sections.

The second key point of no lesser importance is the legal framework. It is clear that examples from the USA like *The Technology-Related Assistance for Persons with Disabilities Act* (1988) and *The Americans with Disabilities Act* (ADA) (1991) are important examples to follow in the European Union and other regions of the world. The publication of the ADA has given a legal tool to private and public organisations in their demands for more accessible technology and they can now require the use of accessible hardware and software by the administration, the school system or universities. The existence of a legal framework cannot be neglected and imposes a certain degree of obligation on the implementer, for which guidelines can be helpful to facilitate his or her design.

### 2.3.3 Destruction of common beliefs

The process of creating awareness goes hand in hand with the process of destroying common beliefs. The authors have met with this issue several times. This section describes some of the arguments found and how to deal with them.

#### *2.3.3.1 Is inclusive design cost-effective?*

The argument of supplementary costs being added to the final product because of the consideration of inclusive guidelines is one of the first arguments used against their implementation—and one of the hardest to destroy. It is evident that there is an initial cost associated with training implementers of guidelines and, of course, associated with their learning curve. There are ways to overcome this argument. The first one is associated with the increase of the market potential for the final product. As a useful example, the 10$^{th}$ WWW User Survey of the Graphic, Visualization & Usability Center (GVU) of Georgia Tech[7] points out that almost

---

[6] "*Websites That Work!*", by the Royal National Institute for the Blind, www.rnib.org.uk/, and the Web Accessibility Initiative, www.w3.org/WAI/

[7] www.cc.gatech.edu/gvu/user_surveys/

10% of Web users have some kind of impairment. It is quite a substantial share of the market and we are sure that implementers cannot afford to neglect this fact.

The second fact is linked with the legal environment of the market. As mentioned earlier, there are countries, e.g. the United States with the Americans with Disabilities Act, where the legal framework imposes the obligation to consider the needs of disabled and older people. The rejection to implement inclusive guidelines can increase the cost of development of the product or service in the long run, because its retrofitting is obviously more expensive.

The average software developer or Web designer uses authoring tools that, however, hide subtleties and technical aspects from the designer. As an example, one of the authors of this chapter has met several times with Web designers who do not know HTML at all. Needless to say, they will hardly be able to implement any recommendation for Web accessibility. To address this issue, the Web Accessibility Initiative (see Chapter 10) has a working group on the topic of authoring tools. The authors recommend that attention is paid to this issue because it can place a big hurdle to the success of guideline implementation.

#### *2.3.3.2 Is inclusive design appealing? Will the product lose functionality?*

This argument has been used for years, especially with regard to Web accessibility. It is very common to associate Web accessibility with boring text-only pages, without any multimedia content. This misconception can be easily extended to the development of any software, and the trainer must stress the positive side of the guidelines. Guidelines are never about forbidding the inclusion of materials but on providing tools and methods to make them accessible for everyone.

This discussion can be taken a step further. Whenever possible, implementers should avoid double versions of products or services (software, Web pages, etc.). The first obstacle will be the cost of maintaining two versions (which has economic implications), but user acceptability is very important as well. It is not uncommon for a product to be rejected because of an external appearance far different from that of mainstream products.

Furthermore, the idea of a mainstream product being used by a wider audience is commercially appealing. The challenge is to develop a common-looking product, which can also be functional to end-users with special needs. The help of guidelines in this context must be underlined, as well as the concept of universal design (Story, 1998; Stephanidis *et al.*, 1999).

Some simple examples will illustrate. Household devices often require the attention of the user, perhaps displayed with a quite small status indicator. Adding an audible signal to the device to indicate that a user action is required, and accompanying it by a small blinking light, makes this product usable for both deaf and blind people, and other users might even consider this as a handy warning!

Another example is the inclusion of accessibility features in some operating systems. This issue was first included in some versions of Mac OS (Apple) and OS/2 (IBM), and shortly afterwards incorporated in Windows95/98/NT (Microsoft). This addition was not considered as being disruptive by the common user and even helped people without a disability—like left-handed people trying to use the mouse, or people having problems following the pointer on the screen.

A third example comes from the Web. It is a well known fact that the use of the *Web Content Accessibility Guidelines* (www.w3.or/TR/WCAG, and see Chapter 10) benefit not only disabled people but users of cutting-edge technology (hand-held computers, Internet telephones, embedded devices, etc.) and facilitate site-indexing by search engines and Internet robots.

#### *2.3.3.3 Why should I care? Assistive technology will overcome the barriers!*

In the past few years assistive technology has provided a set of powerful tools that allow disabled and older people access to information and services. However, there is something that no technology can do: guess the intention of the designer. Human–computer interaction is the link between a product and the user. The implemented interface must be able to provide information—by visual or other means—so that the end-user is able to interact with it. The same happens with technology; it can only provide whatever information is given. If you do not provide an ALT-text tag for an image in a Web site, how could the screen-reader deal with it?

Therefore, the idea of inclusive design is to present the information given by different elements of the interface in some additional ways, thus giving different assistive devices the opportunity to interact with the user with special needs.

### 2.3.4 A training programme with Web designers

This section will describe a training experience with a group of Web designers working for one of the largest telephone companies in Spain. This group is used to designing Web sites of great size and complexity, and they were 'forced' to learn about Web accessibility because of a disability-related Web site.

#### *2.3.4.1 Awareness*

To illustrate different accessibility issues and how to tackle them, they were exposed to a session where they were shown different types of disabilities and how to tackle access to computers. In particular, they found out about:

- alternative input devices: keyboard emulators, pointers, switches, sticks, mouse emulators
- auxiliary devices: guides and holders
- speech recognition software
- speech synthesisers
- software aids.

They then had a 'browser-awareness' session. We showed them different standard browsers, text-based browsers, voice-based browsers and screen-readers. In particular they used:[8]

- Internet Explorer
- Netscape Communicator
- Opera Web Browser
- WebTV
- Emacs_W3
- Lynx
- IBM Home Page Reader
- Sigtuna Browser (English version).

They browsed the Internet with these tools and, in particular, they examined their own designs. This gave them an idea about the consequences of ignoring accessibility guidelines for people with disabilities.

### *2.3.4.2 Training sessions*

The following step was to show them how simple things such as adding ALT-tags to images, including d-links or providing alternatives to scripts, could improve accessibility dramatically. This gave the trainer the opportunity to discard the myth of cost-effectiveness for accessibility. An intensive training followed on how to implement the *Web Content Accessibility Guidelines* 1.0. They learnt—as remarked in the *Guidelines*—that following them will also make Web content more available to all users, whatever user agent they are using (for example, they were not aware of the possibilities for mobile telephone access) or constraints they may be operating under. They learnt as well that we were not discouraging content developers from using multimedia but rather explaining how to make multimedia content more accessible to a wider audience.

### *2.3.4.3 Results*

Shortly afterwards, this group of people began their design and the results of the training were excellent: 80% of the designed pages conformed to the Guidelines to level-A and the rest required only simple changes. Even a level-AA was not difficult to acquire after a revision. Level-A implies that all priority 1 checkpoints of the *Web Content Accessibility Guidelines* are met (otherwise, one or more groups of users would find it *impossible* to access information in the document). Level-AA implies that all priority 2 checkpoints of the *Web Content Accessibility Guidelines* are met (otherwise, one or more groups of users would find it *difficult* to access information in the document; satisfying this checkpoint will remove significant barriers to accessing Web documents).

---

[8] For further references, Browser Watch is a list of browsers compiled by an independent source that can be found in browserwatch.Internet.com/browsers.html

This experience was very positive, and even though it was not clear whether the management of the company would implement these procedures in other design projects, it will surely influence the future work of the designers.

### 2.3.5 An example on software development: Java

We want to finish this section mentioning the advance offered by Java™ to develop accessible software. Java is a growing object-oriented language with applications, not only in PCs or workstations, but in many other devices from mobile telephones to environmental control devices. Its portability, together with its versatility, can ensure a promising future for this technology.

Training software developers of Java-based applications has been eased by efforts from the Sun Accessibility team that enabled the designer to incorporate assistive technology input and output into the software. In fact, recent versions of the Java Development Kit[9] contain the Java Accessibility API (Application Programming Interface), the Swing user-interface classes and support for loading assistive technologies into the Java Virtual Machine. They also developed an Access Bridge to the API of the Microsoft Windows operating system, to be incorporated in assistive technology products. On this token, we cannot forget the efforts of the IBM Special Needs team who made available the *IBM Guidelines for Writing Accessible Applications Using 100% Pure Java.*[10]

## 2.4 THE VERIFICATION/EVALUATION PROCESS

### 2.4.1 Interaction between end-users and implementers

Once end-users and implementers have gone through their training phases, the next step is to establish an interchange phase. We must point out that there are always occasions when this extensive approach is not feasible, many times due to economic reasons. Pros and cons were examined in detail in section 2.2. However, the interaction of the end-user with the product is always desirable, as pointed out by many authors (see, for example, Furugren and Lundman, 1995; Pascoe *et al.*, 1995; and Velasco, 1998) and in the design of mainstream products such as software, Internet sites, interfaces for household equipment, etc. This is where

- guidelines must come into play and
- trainers of guidelines implementers must play several roles simultaneously.

We strongly recommend interaction of end-users with the final product or service whenever feasible. For products addressed to different cultural audiences, Tahkokallio (1998) points out a singular experience that might not be realistic commercially, but, nevertheless, it stresses the benefits of that interchange. This experience refers to a multi-cultural workshop where designers and design students

---

[9] java.sun.com/products/jdk/1.2/

[10] www-3.ibm.com/able/overview.html

from different continents were invited to participate. The conclusions addressed an inherent difficulty to find a common area in the design process, due to cultural differences. However, we feel that every design process is a trade-off, where a minimum usability has to be obtained. An 'ideal' verification/evaluation process should contain the following steps:

#### *2.4.1.1 Implement a design guideline*

Implementers should introduce a design guideline for the product or service under consideration. Whenever feasible, end-users must have access to both versions of the product (with or without the guideline implemented) and must be informed about its expected performance.

#### *2.4.1.2 Test functionality*

End-users must verify and test whether the 'inclusive' version (the one with the guideline implemented) is allowing them to access the complete set of capabilities of the product or service as described to them previously. This phase is not at all trivial. We have mentioned earlier that end-users must be trained; however, they must have access to appropriate tools to perform this task efficiently.

#### *2.4.1.3 Exchange phase*

An exchange forum between end-users and implementers, assisted by means of a definition of common languages in order to facilitate bi-directional communication, must be established. This forum will validate:

- whether any given guideline is fulfilling the expected results from the point of view of implementers *and* end-users, or
- whether any given guideline must be modified, and how, in order to achieve the aims it was intended for.

Each target product, or each guideline, must be verified and tested with the resources available. A positive disposition to collaborate between manufacturers and end-users will benefit both sides. Massive evaluations of guidelines must arise as a result of a strong cooperation between industry and organisations of end-users. As an illustration, we will describe now our experience with a software company and allow the reader to extrapolate from it.

### 2.4.2 An example of collaboration

Since the beginning of 1999, ISdAC has been cooperating closely with a world-wide leading company of speech and language solutions, including software products, services and technologies for speech recognition, language translation, speech synthesis and audio compression. They have developed a wide range of products, which were not specifically designed with the idea of inclusive design in mind but were targeted towards the business world.

When ISdAC found out about one of their products, which allowed users to operate virtually every MS-Windows application by voice, as well as dictate text fluently with a high rate of accuracy, they contacted the company and pointed out the potential market of disabled users for such a product. This resulted in our opportunity to test it.

Through our channels (Web site, forum-list), disabled people were also informed about our experiences. Since then, ISdAC has been testing new releases of the software and was especially very active in the research and development (R&D) process of a local language version.

The company had been using the system of Beta-testing (under a non-disclosure agreement) by a pool of users for some time. They sent out a number of prototypes, accompanied by a set of tests to be conducted. The results, together with a questionnaire, were returned to their R&D department, where final adjustments were made before the product was marketed. The Beta testers were mainly small-to-medium sized enterprises (SMEs), and a few individuals, but it was not until recently that people with disabilities were also added to this group through ISdAC.

When we looked at the experiences of both parties that came out of this cooperation, we noticed a few striking facts. During the different tests that ISdAC was carrying out in this field, it was proven that the feedback given by our organisation was detailed and of high quality. During the test periods, some of the ISdAC members were almost on a daily basis in touch with members of the R&D department by telephone, fax or e-mail, reporting bugs, problems and sometimes even minor details. Simultaneously, ISdAC was keeping the Marketing Department informed about its activities, publications and promotion towards the disabled constituency, where their experiences with the product were shared. This way of working was greatly appreciated by the manufacturer and they thought of the ISdAC feedback as being very valuable. At the moment, ISdAC is considered as their contact with the disabled target group on a European level. Copies of the software are being passed on to ISdAC contacts in Europe to test, evaluate and, if satisfactory, promote on a local scale.

From the user's perspective, we noticed that the ISdAC end-users had the feeling that their efforts were appreciated and their needs were taken into account, and, as a result, they were very motivated to provide a valuable input. Since there was a close contact with the R&D and the Marketing Departments, they also became aware of the fact that sometimes implementing certain guidelines within certain products would imply significant modifications or very high production costs and therefore compromises were necessary. However, both parties agree to aim for a maximum inclusive design approach, so that maximum usability is guaranteed.

## 2.5 CONCLUSIONS

The training, verification and evaluation of guidelines are difficult tasks. There are no magic recipes which can solve problems instantaneously. Our intention is mainly targeted to show the reader practical examples from which others could learn. The ideal situation will arise when interaction between end-users and implementers can take place. However, as pointed out earlier, a successful interaction requires willingness to listen and to take into account the other party's demands. Often this implies giving in on certain views in order to create a workable platform for constructive collaboration.

However, the lack of this interaction does not imply that inclusive design guidelines are impossible to apply. We have shown some methods that will promote successful training of the implementers of guidelines. In any case, the target should always be to collect experiences and to make them available to the community. There is an open field of research on adequate tools and methods to facilitate interaction with end-users which must be explored. This will lead certainly to successful implementation of guidelines in mainstream products and services.

## CHAPTER THREE

# Accessibility and Usability Requirements for ICTs for Disabled and Elderly People: a Functional Classification Approach

Helen Petrie

## 3.1 INTRODUCTION

Information and communication technologies (ICTs), whether they are personal computers, mobile telephones, automatic banking machines (ATMs) or the World Wide Web (WWW), are increasingly becoming an integral part of day-to-day life for business, education and leisure. For disabled and older people to participate fully in these activities, it is vital that all these technologies are made fully accessible and usable by them. Yet for designers working on these technologies for mainstream users, it is difficult to find information about how to make them accessible to a wider range of users. There are a number of sources of general information and sets of recommendations and guidelines for the design of systems for disabled and older users exist (Morris, 1994; Poulson *et al.*, 1996; Sony, 1998; Spiezle, 1999; World Wide Web Consortium, 1999). These sources of information are all useful and meet different needs and situations but still suffer from a number of limitations. The usability of guidelines for designers I am addressing in a separate strand of research (Colwell and Petrie, 1999)—here I will discuss the level of generality and the comprehensiveness of sets of guidelines.

Often guidelines make very general suggestions which relate to very broad categories of disability or ageing. For example, the Sony (1998) guidelines state that "consistent aural cues should be produced in response to operations", which is clearly important and useful for all users who lack vision to see a display. (For simplicity and generality, this chapter will use the term *display* to refer to any presentation of information from an ICT—whether that is a computer or mobile phone screen, the display on an information kiosk or a public transport ticket machine.) However, it does not give the designer much information about how best to design the aural cues and what consistency means. In addition, it is probably more helpful for people using an ICT without vision at all, as it does not lead the designer to think about the different ways that users may have partial vision, the different kinds of visual and auditory information they are likely to be able to perceive and how best to use those capabilities to the full. The other problem with sets of guidelines is that it is not usually clear whether they have covered the full range of disabilities and/or effects of ageing. For example, Spiezle (1999) presents his guidelines for Web design for older users in categories relating to components of Web pages—layout, style, colour, etc. While this is undoubtedly useful for designers when actually working on a specific problem, it does little to increase their understanding of the problems which older Web users actually have. In terms of assessing whether this is an adequate and comprehensive set of guidelines, it is not clear whether all the problems that older users have have been addressed (e.g., older people tend to process cognitively new information more slowly than younger people, which would seem to have interesting implications for Web design, but it is not clear where that would be dealt with in these guidelines).

One approach to addressing the comprehensibility problem might be to look at existing classifications of disabilities to discover whether this provides an outline of the problems associated with each disability that might then be mapped to the requirements for accessing ICTs. For example, the first version of the World Health Organization's International Classification of Impairments, Disabilities and Handicaps (ICIDH) (World Health Organization, 1980) developed a widely used classification of impairments, disabilities and handicaps. The terms 'impairment', 'disability' and 'handicap' are often used interchangeably, however the WHO classification drew useful distinctions between them (see also Stopford, 1987). A particular disease, disorder or pathological condition may lead to an *impairment,* which is:

> . . . concerned with abnormalities of body structure and appearance and with organ or system function, resulting from any cause; in principle impairments represent disturbances at the organ level (World Health Organization, 1980, p. 14)

An impairment, in turn, may lead to a *disability*:

> . . . reflecting the consequences of impairment in terms of functional performance and activity by the individual; disabilities thus represent disturbances at the level of the person (World Health Organization, 1980, p. 14)

Finally, a disability may lead to a *handicap*:

> . . . concerned with the disadvantages experienced by the individual as a result of impairments and disabilities; handicaps thus reflect interaction with and adaptation to the individual's surroundings (World Health Organization, 1980, p. 14)

Thus, an individual may have a *disease* such as diabetes which leads to an *impairment* of functioning of the lens of the eye, known as cataracts. This in turn leads to the *disability* in relation to tasks requiring adequate visual acuity, for example reading, recognition of faces, writing and visual manipulation of objects. This disability results in a number of handicaps, for example of decreased independent mobility and social integration.

More recently, the WHO has begun work on a revision of the ICIDH (World Health Organization, 1999), known as ICIDH-2. This proposes an even more complex classification based on three dimensions: the *Body* dimension, which comprises two classifications, one for functions of body systems and one for the body structure; the *Activities* dimension, which covers the complete range of activities performed by an individual; and the *Participation* dimension, which classifies areas of life in which an individual is involved, has access to, or has society opportunities or barriers. Finally, a list of environmental factors forms part of the new ICIDH-2 classification. Environmental factors have an impact on all three dimensions and are organised from the individual's most immediate environment to the general environment.

The WHO classifications are organised on the basis of a structural division of the body parts and organs which may be affected and also on the severity of the problem. The classification of severity in ICIDH-2 is: complete, severe, moderate, mild, none. The classifications yield little information which can be used to understand the consequences of particular conditions in real life situations in terms of the function, or lack of function, for those who do not already understand the conditions. This is the type of information that could be extremely useful in the design of new technology or the redesign of existing technology for better accessibility for people with these conditions.

Another common approach to the classification of impairments, disabilities and handicaps is to start from the medical conditions which result in impairments (e.g., Stopford, 1987). However, this approach suffers from similar problems to structural classifications such as the ICIDH and ICIDH-2 in that the information is of a very general nature and does not relate directly to how individuals with these conditions undertake real world tasks. Thus it is of little use to a designer without a detailed understanding of the symptoms and behavioural consequences of such symptoms.

In response to this lack of a comprehensive functional survey of disabilities, this chapter proposes a functional classification of the capabilities of the main human information processing channels (vision and audition), human motor capabilities and human cognitive capabilities which are relevant to the design and use of ICTs. By focussing on each of these capabilities, analysing what can malfunction with each capability and then considering the implications of those malfunctions for the use of ICTs, a useful and thorough set of principles for the accessibility and usability of ICTs for disabled and elderly people will be achieved. The analysis in this chapter is only the first step towards such a set of principles.

It is important to note that the emphasis is on usability as well as accessibility. In discussing ICTs for people with disabilities, the emphasis recently has been on accessibility. The assumption has been that if ICTs are made accessible to users with disabilities they will automatically be as usable as they are for non-disabled users. For example, text in an ICT (e.g., on the screen of an ATM or a mobile phone) can be converted to synthetic speech—and many people might consider that is all ICT designers need to do to make it usable by people with visual disabilities. However, without providing appropriate interaction methods for allowing people to navigate, explore and manipulate the information when it is presented in synthetic speech, it will remain unusable, even though visually disabled people can hear it.

To provide a classification of disabilities that will provide useful input to the design situation, I propose a classification based on the *functions* of the different physical, sensory and mental capabilities of human beings and the dimensions along which these functions can vary. For example, one of the functions of our visual system is to perceive detail in the visual scene. This can be considered in terms of a number of dimensions, such as the ability to perceive detail in the foveal area as opposed to in the periphery, the ability to perceive detail for objects at different distances from the eye (i.e., the ability to focus on objects) and so on. By considering the functions of different components of human beings and the dimensions of these functions, one can quickly move to how these functions are used in real world tasks to be undertaken with ICTs and to the consequences of malfunctionings for such real world tasks. In developing this type of classification, one can also consider the incidence of different types of malfunctionings, temporary or permanent, to provide information about the relative importance of different possible design options.

This approach to classification—which I term the functional classification of disabilities—brings with it a number of advantages:

1. It places people with disabilities firmly within a framework of design parameters for all individuals. It emphasises the fact that everyone is disabled at certain times, in certain situations. That may be a momentary disability, such as trying to use a public telephone in the dark. It may be a temporary disability, such as having one's dominant hand in plaster for some weeks after a fracture. Or, it may be a permanent disability from an inherited or acquired disease or trauma or through the natural ageing process.
2. It avoids the need to categorise people with disabilities into groups which mask enormous heterogeneity and individual differences. Groups such as 'partially sighted' or 'hard of hearing' represent people with great variation in their capacities on a number of dimensions of sight and hearing. By considering each of the relevant dimensions individually, we are able to illustrate how they affect people's abilities to undertake real world tasks.
3. By considering the actual dimensions of functional ability and disability, we will eventually be able to provide quantitative information about each of the dimensions, which is impossible for the more heterogeneous groupings, such as 'partially sighted'. This should allow for more targeted design decisions in terms of the importance of different design possibilities.
4. The problems of older people can easily be integrated into the classification or profiles of the common patterns of impairment and disability experienced by elderly people can easily be extracted from the classification, by considering the effect of ageing, both normal and pathological, on the dimensions of each of the functions under consideration.

In developing a functional classification of disabilities, I will consider here the functioning of the visual and auditory systems, motor capabilities and cognitive capabilities as these are the most relevant, or potentially relevant, to the tasks involved in using ICTs. In the following sections I will present the functional classification of the impairments and disabilities of the capabilities of the visual, auditory, motor and cognitive systems.

## 3.2 VISUAL CAPABILITIES, IMPAIRMENTS AND DISABILITIES

The key functional capabilities of the visual system, with their common impairments and disabilities and resulting suggestions for accessibility and usability for ICT design are shown in Table 3.1 and are discussed below. (Tables 3.1–3.5 can be found at the end of this chapter.)

### 3.2.1 Acuity

Perhaps the most basic function of the visual system is to distinguish patterns of light and dark as figure and ground (object and background) in the world. Various diseases, congenital conditions and the effects of ageing affect the ability to make this basic visual discrimination. Some conditions result in complete loss of basic vision, although the percentage of people affected in this way is relatively small. For example, retrolental fibroplasia, which is thought to be caused by excessive oxygen at birth (Hollins, 1989) causes total blindness. Cataracts, which account for about 10–15% of all cases of blindness in the developed world (Hollins, 1989), result in a loss of basic vision, which may be total or partial.

A number of factors affect this ability which can be used to maximise the perception of objects in real world situations. One factor is brightness contrast: making strong contrasts between the figure and the ground, for example between buttons and the surrounding surface or between screen objects and the background. The other factor is perceptual grouping: objects should have clear, well-defined boundaries and objects which should be grouped together should have clear organising principles.

Beyond the perception of objects and backgrounds, one of the most fundamental visual capabilities is the ability to perceive detail in those objects and backgrounds, known as acuity. Acuity is the function measured by standard eye tests and rated by the Snellen Scale.[1] Static acuity is the ability to perceive detail in stationary objects, which might include information on smart cards, or information on a terminal screen or keyboard, be that text or graphic information. Some diseases such as cataracts cause loss of acuity throughout the visual field, whereas others such as macular degeneration cause loss of acuity only in the most sensitive area of the retina (the macula), so the person can see the periphery but not the centre of their visual field.

One of the most important tasks for which reasonable acuity is required is reading. Unfortunately, it has been found that acuity as measured by the Snellen

[1] The Snellen Scale is the standard measure of visual acuity and the standard for defining blindness. It is typically measured with an eyechart, placed 6 metres (or 20 feet in the United States) away from the person being tested. The average person with unimpaired vision can just read letters of a certain size at this distance and is said to have 6/6 vision (or 20/20 vision in the USA). A person who cannot read these letters, but is just able to make out others of twice that size, has 6/12 vision, which is to say that he or she can just read at 6 metres letters that someone with unimpaired vision could distinguish at a distance of 12 metres from the chart. The worse a person's acuity, the larger the denominator in the sight rating. Someone who, even when wearing glasses, has an acuity of of 6/60 or worse—that is, 10 times poorer than that of a person with unimpaired vision—is legally blind in the USA.

Scale has little relationship to the performance of everyday reading tasks (Legge *et al.*, 1985), so one cannot make simple recommendations such as progressively increasing print size to compensate for poorer acuity. However, several points have emerged from research which can guide designers. Magnification of text does help people with poor acuity to read, but magnification beyond 3 to 6 degrees does not improve reading performance. Likewise, increasing the character spacing beyond values used in normal text does not improve reading performance. It should also be remembered that reading times for people with poor acuity can be very much longer than for people with good acuity.

Acuity varies with illumination level and various visual impairments interact with the normal range of acuity in different illumination levels. For example, some conditions make it difficult to perceive detail in low illumination levels. Acuity also decreases in high illumination levels, which we all experience as glare, and the ability to perceive detail in glare decreases with age.

Dynamic acuity is the ability to perceive detail in moving objects. This may not seem very important for current terminal technology but as multimedia systems become more common and include animation, video clips and other moving objects, the ability to understand objects as they move will become more important. The disabilities associated with static acuity tend to be magnified for dynamic acuity, with perception of detail in moving objects even more difficult for people with poor acuity.

### 3.2.2 Accommodation

Accommodation is the ability to focus on objects at different distances from the eye. We are all aware of the problems of shortsightedness (myopia) and longsightedness (hypermetropia) and the fact that these can often be corrected by wearing spectacles. Unfortunately, whereas when reading from a portable device such as a book or paper, one can move the device to an appropriate distance from the eyes to optimise the distance for focusing, this is not possible with a conventional terminal. This problem is exacerbated by the fact that terminals tend to be at a greater distance than is normal for reading hardcopy text (Sanders and McCormick, 1992), so corrective spectacles may not assist a person in focusing at the appropriate point.

### 3.2.3 Adaptation

Adaptation is the way the eye changes its capabilities between very low illumination levels and normal illumination levels, due to the presence of two types of cells in the retina, cones and rods. The process of adaptation is noticeable when one enters a dark environment such as a cinema—at first nothing is visible apart from what is on the highly illuminated screen. Over a period of time, the eyes *dark adapt* to the low illumination and one sees dimmer and dimmer objects such as the empty seats and faces of surrounding people. The opposite effect occurs when leaving the cinema—initially it is difficult to see anything outside the cinema, as the eyes are not adapted to the high-illumination conditions. Various conditions such as xerophthalmia (although this is rare in developed countries) result in loss of vision in dark-adapted situations, known as night blindness. This even occurs significantly in student populations who eat poorly, as it is thought to be related to vitamin A deficiency (Ripps, 1982). Its most important implication is for driving at night but it may also be relevant to the design of lighting for equipment which is used at night.

### 3.2.4 Perception of depth

The perception of depth in the visual field is a complex process, depending on the use of a number of types of information. Individuals who lack one type of information seem to be able to make up for this deficit by using other types of information and therefore do not necessarily lack depth perception as one might predict (for example, people with strabismus). However, people with severe visual impairment in one eye only may have difficulty with depth perception, as many of the sources of information come from binocular visual cues.

Elderly people also find it more difficult to integrate these sources of information rapidly and may have problems with perceiving depth. This may not be a problem in itself in understanding the two dimensional properties of a card or a terminal screen, but depth perception is vital in co-ordinating one's interaction with the world and may lead to problems of eye–hand co-ordination when using ICT equipment.

### 3.2.5 Colour vision

Approximately 8% of men and 0.05% of women have some form of colour blindness (Alpern, 1981). Unfortunately, there are five different types of colour blindness with different consequences for the colour discriminations that are possible. The most common form is deuteranopia, also known as red-green deficiency, in which light in the 'green' area of the wavelength is responded to but green cannot be distinguished from certain combinations of red and blue. Therefore colour contrasts between violet and blue or within the green/yellow/orange/red segment of the colour spectrum should be avoided.

Perception of colour is also affected by the general ageing process and a number of diseases and disorders such as diabetes and glaucoma. In all these cases, the loss of colour perception is towards the violet/blue end of the colour spectrum. Therefore fine contrasts at this end of the colour spectrum should be avoided.

## 3.3 AUDITORY CAPABILITIES, IMPAIRMENTS AND DISABILITIES

The key functional capabilities of the auditory system, with their common impairments and disabilities and suggestions for guidelines for accessible ICT design, are shown in Table 3.2.

### 3.3.1 Frequency

Human beings can generally hear sounds in the frequency range of approximately 20 hertz to 20,000 hertz.[2] The physical characteristic of frequency is generally related to the psychological dimension of pitch, with higher frequency sounds being higher pitched. However, the intensity of a sound (see Section 3.3.2 below) will also affect its perceived pitch; the relationship is complex, with low-frequency sounds sounding lower in pitch as intensity increases but high-frequency sounds sounding higher in pitch as intensity increases.

Some forms of deafness, particularly conductive deafness, make sounds of all frequencies more difficult to hear. For people with partial deafness of the conductive variety, all sounds need to be more intense to be heard. Other forms of deafness and the effects of ageing lead to the loss of perception of sound of a particular frequency range. For Ménière's disease this tends to be low-frequency sounds, whereas the general effect of ageing and other pathological conditions lead to the progressive loss of perception of high-frequency sounds. Therefore it is impossible to predict which range is critical for all deaf people, although it could be that an ICT system could have information stored about the range of frequency loss for particular individuals.

A further important psychological function related to frequency is the ability to discriminate between the frequency of two sounds, or pitch perception. For example, if two different tones were used to indicate different states of a terminal, pitch perception would be important. Individuals vary greatly in their ability to discriminate between sounds of different frequency or pitch but this ability generally decreases with age. Therefore if such discriminations are important, they need to be large discriminations which all users are likely to be able to make.

### 3.3.2 Intensity

The intensity of a sound relates to the amplitude of the sound wave. In general terms, the greater the amplitude or intensity of the sound, the louder it is perceived by the human ear. Humans can hear sounds up to a loudness of approximately 150 decibels,[3] although prolonged exposure to sounds over approximately 100 decibels can lead to hearing loss.

With numerous forms of deafness and the general effect of ageing on the auditory system, sounds must be at a higher intensity to be heard. Some impairments of the auditory system can also lead to distortions of the perception of loudness, known as 'recruitment'. Low-intensity sounds may not be heard at all, but sounds of higher intensity are heard as much louder than normal and may be painful. Thus having louder sounds for all deaf users would be inappropriate and even counterproductive.

---

[2] Hertz is the unit for measuring wave frequencies, 1 hertz being equivalent to one wave cycle per second.

[3] The decibel is the standard unit of measurement for sound wave amplitude. The decibel is a logarithmic scale derived from the ratio of the air pressure amplitude of a sound in relation to a standard level.

### 3.3.3 Signal-to-noise ratio

Listening to sounds in the real world involves not simply hearing an individual sound against a quiet background but picking out the relevant sounds from all the 'noise' going on around. If the ratio of relevant sound to noise is poor, important information will not be received. One of the general effects of the ageing of the auditory system is decreasing ability to hear in a noisy environment. Paradoxically in some forms of deafness, such as otosclerosis, it may be easier to hear in a noisy environment, although this may be an artifact of people speaking louder in noisy environments. Thus the background noise environment of any system needs to be considered in the ICT design process.

## 3.4 MOTOR CAPABILITIES, IMPAIRMENTS AND DISABILITIES

The key functional capabilities of the motor system, with their common impairments and disabilities and suggestions for guidelines for accessible ICT design, are shown in Table 3.3.

For the current purposes, the functional capabilities of the motor system can be divided into two areas—those associated with the lower limbs and those associated with the upper limbs.

Any malfunctioning of the lower limbs may result in the need to use a support for walking such as a stick, crutches or walking frame, or if the malfunction is severe enough, the person may need to use a wheelchair. Current terminals can be accessed by people using walking supports, but if the kinds of transactions which users perform with terminals become more complicated and time-consuming, it may be important to consider the need for terminals with seats for users.

People in wheelchairs already have difficulty with access to public ICT terminals such at automatic teller machines (ATMs) and ticket machines, in terms of height and reach (see also Chapter 14). For example, the typical floor-to-shoulder height for a male wheelchair user is 108 cm and for a female wheelchair user 98 cm (these and subsequent statistics from Pheasant, 1986). Therefore the height of insertion and retrieval slots at public terminals should not be greater than 100 cm. Similarly, the typical floor-to-elbow height for a male wheelchair user is 68 cm and for a female wheelchair user 67 cm. Therefore the height of keypads should not be greater than 67 cm. In terms of reach, wheelchair users need either to access a public terminal straight on with a cavity under the display to accommodate the lower portion of their body or to approach a terminal sideways, although this option would involve twisting the body round to use the terminal which could create difficulties. For a straight-on access arrangement, the height of the cavity would need to be at least 74 cm and the depth at least 44 cm to accommodate average wheelchairs. For a sideways access arrangement, the typical sideways reach for a male wheelchair user is 64 cm from the edge of the wheelchair and for a typical female wheelchair user 58 cm. So distances to the terminal would need to be within within 58 cm.

The capabilities of the upper limbs are rather more complex in relation to ICT design. Firstly, an individual may lack all function of one or both of their upper limbs due to the lack of the limbs or paralysis of the limbs. Limbs may also be incapable of making fine movements because of swelling and stiffness in the joints, as found with arthritis. Rheumatoid arthritis affects people of all ages, whereas osteoarthritis is caused by general ageing of the joints, so is common amongst the elderly.

For some actions required in ICT use, alternatives to upper limb actions are possible. For example, information normally entered via a keyboard or keypad can be entered via voice input. Contactless smart cards can resolve the problem of having to insert a card into a slot. For receiving items from a public terminal such as cash, receipts or tickets, a system where information is stored on the card itself has obvious advantages.

Other dimensions of the functionality of the upper limbs are the capability to co-ordinate and control actions, which is lost in some conditions such as multiple sclerosis. Controlling actions also require hand–eye co-ordination, so sufficient visual capabilities are also required. Lack of control of limb movements may also be due to tremors, as found in Parkinson's Disease. In these cases the alternatives mentioned above will alleviate the situation, but it may also be that designing the terminal so that very fine controlled movements are not critical will assist, for example, by having large keys and buttons to press so that fine motor control is not as critical.

## 3.5 COGNITIVE CAPABILITIES, IMPAIRMENTS AND DISABILITIES

The key functional capabilities of the cognitive system, with their common impairments and disabilities and suggestions for guidelines for accessible ICT design, are shown in Table 3.4.

The capabilities of the cognitive system are related to each other in complex ways, which means that malfunctions in one dimension are often associated with problems in other areas. For example, an individual with short-term memory problems will have problems with concentration, planning and learning, as these all require short-term memory. However, it was thought useful to list a number of complex cognitive skills in addition to the fundamental ones to see their effect on ICT design.

Memory capabilities, both long term and short term, are important for all mental tasks undertaken. Long-term memory problems will affect factors such as the capability to remember PIN numbers and the functions of particular keys or buttons. Recognition memory, where there is some cue to the information, is always better than recall memory, where no cues are present. However, in the case of secure information such as PIN numbers, providing cues may not be possible. For elderly people, long-term memory is often better for events from early life, so PIN numbers might be replaced by passwords (incorporating numbers) related to the earlier life events of the user.

Short-term memory is the capability to hold information in consciousness for about 20 seconds if there has been no rehearsal of the information. The capacity of the short-term memory store is well known at approximately seven chunks of information (Miller, 1956), although the chunks can be quite complex units. If any distraction from a task occurs, the contents of short-term memory are usually lost.

Various conditions result in problems with holding information in short-term memory and in transferring it to long-term memory. So information which a user may be expected to remember momentarily—say between different screens in an interaction—would be better if it were displayed continuously for reference. For example, some ICT systems give the exit command on the first screen of information and then expect users to remember this for use at the end of the interaction. Continuous display of the exit command would be much more usable.

The speed with which people can process information is a further important dimension of cognitive capabilities. A number of conditions, such as dyslexia and the general effects of ageing, tend to slow the rate of information processing. This is particularly important in the design of time-critical sequences such as the time allowed to enter a PIN number or password. However, it is also important in the design of the presentation of all information which needs to be processed, for example, all displays of information which require comprehension and decision-making such as the choice of a type of ticket to purchase.

Attention or concentration is another dimension of cognitive capability, which in turn has several components. Attentional span is related to short-term memory and refers to how long a person can concentrate on a task and hold information in short-term store. Intensity of attention refers to how easily a person is distracted from a task they are attempting to complete. Some individuals are much more easily distracted, which might occur either because of information irrelevant to the task being presented simultaneously or by their surroundings.

One of the basic tools for conducting mental work is language, which may be written or spoken. People can have difficulties with the processing of language, which may affect either the comprehension or the production of language (and not necessarily both processes). Particular conditions such as the effects of cerebral haemorrhages can also affect the capability to process either only spoken or written language or both. This clearly has implications for the use of text and spoken messages in interactions and the need for users to enter textual or spoken commands. Non-textual alternatives may be available, although care needs to be taken that these are easily understood and remembered.

Using ICTs often requires planning of a task and a sequence of actions. These may seem simple enough but may cause difficulties to users who have difficulties formulating plans. Options to guide users more carefully through the sequence of actions required to achieve a particular task may be useful here, or the possibility of having the appropriate information already stored in the system or on a smart card.

## 3.6 IMPAIRMENTS AND DISABILITIES ASSOCIATED WITH AGEING

As with the problems of classifying the characteristics and requirements of groups of people with disabilities, it must be emphasised that elderly people are very varied and classification is extremely difficult (Small, 1987; Hooyman and Kiyak, 1988). Younger people tend to view elderly people as similar, whereas in reality it is the 18–35 year old group who are most similar in their abilities (Koucelik, 1982). However, there are a number of well-established tendencies in physical and cognitive abilities which occur with ageing which are relevant to ICT design, which will be outlined in this section and are summarised in Table 5.

First, some definitions of what constitutes old age need to be considered. Most researchers regard 55 years as the beginning of old age, although as the researchers themselves age they may wish to advance this criterion. However, clearly there are enormous differences in the abilities and problems of people aged approximately 55 years and those aged approximately 90 years. Therefore, researchers tend to divide the possible age range of elderly people into either two or three groups:

'young-old' (ages 55 to 75) and 'old-old' (over 75) (e.g., Neugarten, 1974)
or
'young-old' (ages 65 to 74), 'old-old' (ages 75 to 85) and 'oldest-old' (over age 85) (e.g., Riley and Riley, 1986).

### 3.6.1 Visual problems associated with ageing

The numerous variations with age in vision are probably the most commonly recognised age-related changes and have been measured extensively.

Static visual acuity, the ability to perceive detail, decreases steadily from middle age, with an average 80% loss in acuity by the 'oldest-old' group in comparison to 40 year olds (Hollins, 1989). Senile macular degeneration, which affects the ability of the central area of the retina (the macular) to detect fine detail, is the most common cause of new cases of blindness in people over the age of 65 in developed countries (Hollins, 1989). However, in general, loss of acuity is greatest around the edges of the visual field, so perception is less accurate than straight in front of the person (Welford, 1985). These problems are greatest in low-illumination conditions.

Problems with acuity will affect the ability to perceive any detail on an ICT display, for example to read small print or diagrams and to discriminate between icons or pictograms. The situation can obviously be improved by larger print and graphics but also by increasing the contrast between the target and its surround and by increasing the overall illumination of the display. However, too much illumination results in glare, which decreases acuity for everyone. Sensitivity to glare increases with age, so levels of illumination should not be too high.

Elderly people also require longer to perceive detail in visual information. The consequences of such slowing down of perception will be treated with the general slowing down of behaviour (see below).

Dynamic visual acuity—the ability to perceive movement—also declines with age and the effect is even more dramatic than static acuity. Hence, if moving elements are included in a display, greater time must be allowed for elderly people to perceive them.

The best recognised age-related visual change is the decrease in the ability to focus clearly on objects at different distances. This problem is largely overcome by wearing spectacles, although problems remain because, even with spectacles, it is only over a restricted range that objects are clearly in focus.

The accuracy of depth perception also decreases with age (Welford, 1985). This may lead to confusion in the location of objects and apparent clumsiness, or loss of eye–hand co-ordination, which is due to misjudging distances when manipulating objects. This difficulty would result in less accurate and slower insertion of cards into slots and removal of banknotes, tickets or receipts from ICT public terminals.

The sensitivity of colour vision decreases with age, with a general loss of sensitivity across the entire colour spectrum but with particular loss at the blue-violet end of the spectrum. Amongst 60 to 70 year olds, Gilbert (1967) found accurate colour discriminations from 76% of a young adult control group, and amongst 80 to 90 year olds, from only 56% of the control group.

### 3.6.2 Auditory impairments associated with ageing

As with visual capabilities, auditory changes are among the most commonly occurring age-related changes. Hearing is, in fact, one of the major problems for many elderly people. Sounds of higher frequencies become progressively more difficult to detect and require higher intensity thresholds. Discriminations in pitch also become more difficult, particularly for higher-frequency sounds. These factors mean that care needs to be exercised in using high-pitched sounds and discriminations between such sounds.

With increasing age, problems with the detection of relevant sounds against the background of noise also increase and particularly with the perception of complex sound patterns such as speech. This means that speech messages need to be very clear and the surrounding environment ought to be as quiet as possible.

### 3.6.3 Motor impairments associated with ageing

Probably the most prevalent and significant change with age is the slowing of behaviour (Small, 1987) and this has obvious consequences for the use of smart card and terminal technology. This applies not only to relatively simple sensory and motor processes but also to more complex processes involving higher mental operations such as planning activities (Welford, 1985). Many studies with elderly people indicate an increase of at least 20% in reaction time for simple tasks between the ages of 20 and 60 years, with progressive increases for more elderly people.

Problems with the co-ordination and control of movements also increase with age, with implications for the manipulation of small objects such as cards, money, receipts and tickets. Muscle strength also declines with age, meaning that actions that require strength become more difficult.

Problems with the lower limbs mean that many elderly people require support in walking from a walking stick, frame or wheelchair. The problems of access to public terminals for wheelchair users have been discussed in section 3.4 above.

### 3.6.4 Cognitive impairments associated with ageing

Schaie (1983), in summarising many studies involving age changes in cognitive capabilities concludes that those changes amongst the 'young-old' are trivial but that by 'old-old' age reliable decrements do occur. In terms of the magnitude of these changes, he notes that, prior to age 60, no decrement in excess of 0.2 of a population standard deviation in scores can be observed, whereas by age 81, the magnitude of the decrement is approximately one standard deviation for most variables.

Long-term memory retention in elderly people for information learnt one to three decades earlier is good. However, the speed of retrieval from long-term memory depends on the degree of familiarity with what is to be recalled and the compatibility existing in the relationship between the stimuli involved, responses involved, or a combination of stimuli and responses. These relationships derive from a number of factors such as learning and cultural experiences, so what appears compatible for younger people may not be the same for older people. Studies show that lack of compatibility increases task completion times as a function of age. Some evidence also indicates a tendency for errors to occur more frequently with low compatibility and increasingly with age.

As has been noted several times, processing speed decreases with age and this must be considered in relation to both time-critical tasks and the general rate of presenting information.

## 3.7 CONCLUSIONS

This chapter has outlined some of the limitations of previous classifications of impairments, disabilities and handicaps as a source of information for the design of new technologies which will assist disabled and elderly people in their daily living. A classification based on an analysis of the functional capabilities of various human systems has been proposed and has been elaborated for a number of areas relevant to the use and design of ICTs: visual capabilities, auditory capabilities, motor capabilities and cognitive capabilities.

The functional classification of capabilities in these areas has been used to develop suggestions for the design of ICTs which will both make these technologies more accessible to disabled and elderly people and use these technologies to improve the quality of life for such people. The use of the classification in this manner appears to have been initially successful, in that it generated numerous ideas for the design and redesign of ICTs. However, the real value will be seen when it is given to designers in the field to guide their real world design processes.

As with numerous other studies, it has been found that the design of technology to meet the requirements of disabled and elderly people will often be good design for everyone. For example, to make displays easier to use by visually disabled people, clear contrasts between objects and backgrounds are recommended with clear grouping of objects which are related to each other. These are features that make the displays easier for everyone to read. So many of the principles which have been derived from the classification are principles which will make the technology usable by everyone.

**Table 3.1** Functional classification of visual capabilities with common causes of malfunctioning and suggestions for guidelines for accessible ICT design

| Physical capability | Psychological function | Common causes of malfunctioning | Common effects of ageing | Suggestions for guidelines |
|---|---|---|---|---|
| ACUITY | discriminate objects | cataracts<br><br>retrolental fibroplasia | general decrease with age<br><br>80% loss of acuity between age 40 and 85 | (i) Use good contrast between object and background and between different objects. Ensure that apparent brightness difference between objects and background is high but not so high that it creates glare.<br>Some individuals see better when the contrast is 'reversed' from the normal (dark characters on bright background, as in dark ink-print on a white page), so provide facility to reverse contrast (light characters/objects on a dark background). |
| | | | | (ii) Group objects clearly.<br>Place clearly contrasting borders around groups of objects (for contrast, see (i) above).<br>Place grouped objects clearly together and separate from other groups—make the mean distance between centres of groups at least three times the mean distance between the objects within a group. |

**Table 3.1 (cont.)** Functional classification of visual capabilities with common causes of malfunctioning and suggestions for guidelines for accessible ICT design

| Physical capability | Psychological function | Common causes of malfunctioning | Common effects of ageing | Suggestions for guidelines |
|---|---|---|---|---|
| Static acuity | ability to see the world as a continuous field, with peripheral vision of nearly 180 degrees | night blindness<br><br>retinosis pigmentosa<br><br>glaucoma<br><br>diabetic retinopathy | refer to previous section of the table<br><br>acuity loss worse at limits of illumination levels | Solution: vary print and icon size.<br>Be able to vary print font: Helvetica, Arial or Futura should always be available.<br><br>Be able to vary font size: from 6 to 48 point, in continuous, not stepped, increments.<br><br>Icon and button sizes should be variable: from 5 mm diameter to whole screen, in continuous, not stepped, increments. |
| Affected by different illumination levels | | | | Ensure adequate illumination level (minimum 1000 lux), but not excessive illumination such that it causes glare (contrast ratio for adjacent to the centre of the visual field should not exceed 3:1).<br>The contrast ratio between the centre of the visual field and its periphery, or between different objects in the periphery, should not exceed 10:1. |

**Table 3.1 (cont.)** Functional classification of visual capabilities with common causes of malfunctioning and suggestions for guidelines for accessible ICT design

| Physical capability | Psychological function | Common causes of malfunctioning | Common effects of ageing | Suggestions for guidelines |
|---|---|---|---|---|
| For different parts of the visual field | | | | Provide the following commands to change screen position of essential objects/elements:<br>set object/element to centre,<br>set object/element to left/right/top/bottom. |
| Dynamic acuity | discriminate detail in moving objects | as in first section of the table, but effects magnified | as in first section of the table, but effects magnified | Provide facility to slow down moving objects in continuous, not stepped, increments.<br>Provide facility to freeze/restart movement of objects.<br>Provide longer time for people to understand moving objects. |
| ACCOMMO-DATION | ability to focus on objects at different distances | farsightedness (hypermetropia)<br><br>nearsightedness (myopia) | difficulty with focussing on objects at close distances (presbyopia) | Provide facility to move display in a circle with a base of radius 1 metre, in a continuous, not stepped, movement, navigable on complete circumference from base.<br>Provide facility to tilt display bottom up and down to 30 degrees from the horizontal, in a continuous, not stepped, movement. |

**Table 3.1 (cont.)** Functional classification of visual capabilities with common causes of malfunctioning and suggestions for guidelines for accessible ICT design

| Physical capability | Psychological function | Common causes of malfunctioning | Common effects of ageing | Suggestions for guidelines |
|---|---|---|---|---|
| ADAPTATION | ability to see when changing from high illumination to low illumination or vice versa | xerophthalmia<br>night blindness | ability decreases particularly from age 60 | Provide facility to change contrast settings from within the application. |
| COLOUR VISION | Ability to distinguish between the standard colour set | Colour blindness, 5 different varieties with different discrimination losses | Colour vision fades, particularly for blue-violet end of colour spectrum | Contrasts based on violet/blue or green/yellow/orange/red will be more difficult to perceive.<br>Provide full spectrum of colours for both foreground and background in continuous, not stepped, increments.<br>Avoid contrasts based on blue versus green, yellow or orange versus white.<br>Provide facility to change colour settings from within application. |

**Table 3.2** Functional classification of auditory capabilities with common causes of malfunctioning and suggestions for guidelines for accessible ICT design

| Physical capability | Psychological function | Common causes of malfunctioning | Common effects of ageing | Suggestions for guidelines |
|---|---|---|---|---|
| Perception of frequency | ability to hear sound of particular pitch<br><br>ability to discriminate between sounds of different pitches | impairment of a specific frequency range:<br>Ménière's disease (low-frequency loss) | high loss, need louder sound to hear high-pitched sounds<br><br>decreases with age | Pitch discriminations should be large. |
| Perception of intensity | ability to hear sound at particular volumes | nerve deafness: 'recruitment'—quiet sounds lost, loud sounds perceived as louder, painful<br><br>many forms of deafness, e.g. otosclerosis, maternal rubella, infections of middle ear | generally higher thresholds | Sounds should be of moderate loudness, adjustable by the user. |
| Perception of sound in signal-to-noise ratio | ability to hear over background noise | in otosclerosis, may be easier to hear in noisy environments<br><br>tinnitus—creates inner noise background | | Effects of background noise need to be considered. |

**Table 3.2 (cont.)** Functional classification of auditory capabilities with common causes of malfunctioning and suggestions for guidelines for accessible ICT design

| Physical capability | Psychological function | Common causes of malfunctioning | Common effects of ageing | Suggestions for guidelines |
| --- | --- | --- | --- | --- |
| Perception of complexity/ harmony | ability to perceive sounds of different complexities, e.g. pure tones (simple), speech (complex) | otosclerosis | speech perception decreases with age | Distinctions between sounds should be large. |

**Table 3.3** Functional classification of motor capabilities with common causes of malfunctioning and suggestions for guidelines for accessible ICT design

| Physical dimension | Common causes of malfunctioning | Common effects of ageing | Suggestions for guidelines |
|---|---|---|---|
| **Upper limbs** | | | |
| Lack of limb/s | amputation | | In case of inability to use keyboard, provide other input devices; need ability to use voice input, switch control devices. |
| Total paralysis<br>Partial paralysis | spinal cord injury<br>poliomyelitis<br>cerebral haemorrhage | cerebral haemorrhage | In case of total or partial inability to use keyboard, provide other input devices; need ability to use voice input, switch control devices. |
| Lack of co-ordination (ataxia) | multiple sclerosis<br>myasthenia gravis<br>Friedreich's ataxia | | Inability to co-ordinate fine movements such as pointing at icons; provide alternative actions which require little fine control. |

**Table 3.3 (cont.)** Functional classification of motor capabilities with common causes of malfunctioning and suggestions for guidelines for accessible ICT design

| Physical dimension | Common causes of malfunctioning | Common effects of ageing | Suggestions for guidelines |
|---|---|---|---|
| **Upper limbs (cont.)** | | | |
| Tremor | Parkinson's disease<br>multiple sclerosis | | Inability to co-ordinate fine movements such as pointing at icons;<br>provide alternative actions which require little fine control. |
| Joint swelling/stiffness | rheumatoid arthritis | osteoarthritis, caused by general wear and tear of joints due to ageing | Difficulty pressing keys; provide alternative actions which require little fine control. |
| Lack of strength | muscular dystrophy<br>myasthenia gravis | | Difficulty pressing buttons; provide alternative actions which require little strength. |
| Lack of sensitivity | diabetes<br>spina bifida | | Difficulty pressing buttons; provide alternative actions not requiring particular sensitivity. |
| Susceptibility to trauma | Brittle bone disease<br>haemophilia | | Difficulty pressing buttons, manipulating controls; provide alternative actions not requiring particular sensitivity. |

**Table 3.3 (cont.)** Functional classification of motor capabilities with common causes of malfunctioning and suggestions for guidelines for accessible ICT design

| Physical dimension | Common causes of malfunctioning | Common effects of ageing | Suggestions for guidelines |
|---|---|---|---|
| **Lower limbs/trunk** | | | |
| Lack of limb/s | many causes: | | Provide access to terminals for wheelchair users (see section 3.4 for details). |
| Paralysis | spinal cord injury | | |
| Lack of co-ordination | poliomyelitis | | |
| Tremor | arthritis | | |
| Joint swelling/stiffness | | | |

**Table 3.4** Functional classification of cognitive capabilities with common causes of malfunctioning and suggestions for guidelines for accessible ICT design

| Psychological capability | Common causes of malfunctioning | Common effects of ageing | Suggestions for guidelines |
| --- | --- | --- | --- |
| Long-term memory processing | | Alzheimer's disease | Provide simple methods for remembering passwords, PIN numbers, complex instructions. |
| Short-term memory processing | Huntingdon's chorea | Korsakoff's psychosis | Provide continuous feedback about position in the system, task structure. |
| Information processing speed | dyslexia | slows with age | Provide longer to complete any time-critical task. |
| Attention/concentration | Down's syndrome | Pick's disease | Provide continuous feedback about position in the system, task structure. |

**Table 3.4 (cont.)** Functional classification of cognitive capabilities with common causes of malfunctioning and suggestions for guidelines for accessible ICT design

| Psychological capability | Common causes of malfunctioning | Common effects of ageing | Suggestions for guidelines |
|---|---|---|---|
| General level of intellectual processing | Down's syndrome | | Provide alternative simple instructions, with non-orthographic symbols. |
| Language comprehension<br>written<br>spoken | cerebral haemorrage<br><br>dyslexia | | Provide alternative simple instructions, with non-orthographic symbols. |
| Language production<br>written<br>spoken | cerebral haemorrage<br><br>dyslexia | | Provide voice input as an alternative to keyboard input. |
| Planning complex actions | Down's syndrome | | Provide online support for planning the task. |
| Confusions | | Korsakoff's psychosis | Provide continuous feedback about position in the system, task structure, alternatives available at any point. |

**Table 3.5** Impairments and disabilities associated with ageing and suggestions for guidelines for accessible ICT design

| Physical capability | Psychological function | Common effects of ageing | Suggestions for guidelines |
|---|---|---|---|
| **Visual system** | | | |
| Acuity | discriminate objects, detail in objects | general decrease with age | Use good contrast to make objects clearly distinguishable. |
| in different parts of visual field | | macular degeneration causes loss of acuity in centre of visual field, otherwise loss of acuity around periphery of field is more common | Vary print and object size. |
| in different illumination levels | | acuity at extremes of illumination decreases with age | Pay attention to lighting levels. |
| dynamic acuity | discriminate detail in moving objects | decreases with age | Provide ability to slow down moving displays. |
| Accommodation | ability to focus on objects at different distances | decreases with age, particularly for close objects (presbyopia) | Provide displays with adjustable distance. |

**Table 3.5 (cont.)** Impairments and disabilities associated with ageing and suggestions for guidelines for accessible ICT design

| Physical capability | Psychological function | Common effects of ageing | Suggestions for guidelines |
|---|---|---|---|
| **Visual system (cont.)** | | | |
| Depth perception | ability to see how far away objects are | general decrease in accuracy with age | Avoid actions which require fine-grained eye–hand co-ordination. |
| Colour vision | ability to distinguish between the standard colour set | colour vision fades with age, particularly for the blue-violet end of the colour spectrum | Avoid contrasts based on violet/blue, subtle colour differences. |

**Table 3.5 (cont.)** Impairments and disabilities associated with ageing and suggestions for guidelines for accessible ICT design

| Physical capability | Psychological function | Common effects of ageing | Suggestions for guidelines |
|---|---|---|---|
| **Auditory system** | | | |
| Frequency | perceived pitch of sound | | |
| | ability to hear sound of a particular pitch | difficulties with sounds over 1000 Hz | Avoid high-pitched pure tones. |
| | ability to discriminate between sounds of different pitch | ability decreases with age, particularly for high-pitched sounds | Avoid discriminations between high-pitched sounds. |
| Intensity | perceived volume of sound | increased hearing thresholds, increasing for higher-frequency sounds | Avoid soft sounds. |
| Signal-to-noise ratio | ability to hear over background noise | generally decreases with age | Provide quiet environment for terminals. |
| Complexity | ability to understand complex sounds (e.g., speech) | generally decreases with age | Make complex sounds such as speech very clear. |

**Table 3.5 (cont.)** Impairments and disabilities associated with ageing and suggestions for guidelines for accessible ICT design

| Physical capability | Psychological function | Common effects of ageing | Suggestions for guidelines |
|---|---|---|---|
| **Motor system** | | | |
| Upper limbs | Not applicable | | |
| lack of co-ordination | | also related to poorer vision and problems of eye–hand co-ordination | Difficulties co-ordinating fine movements. |
| tremor | | due to weaker muscle strength | Difficulties with fine control actions. |
| stiffness/swelling of joints | | decrease in flexibility due to wear and tear of joints | Difficulties with manipulating small objects such as cards. |
| lack of strength | | due to weaker muscle strength | Difficulties with action requiring strength. |

**Table 3.5 (cont.)** Impairments and disabilities associated with ageing and suggestions for guidelines for accessible ICT design

| Physical capability | Psychological function | Common effects of ageing | Suggestions for guidelines |
|---|---|---|---|
| **Motor system (cont.)** | | | |
| Lower limbs<br><br>lack of co-ordination<br><br>tremor<br><br>stiffness/swelling of joints<br><br>lack of strength | Not applicable | as above | Walking support: stick, crutches, walking frame; may require support, seating for longer transactions.<br><br>Wheelchair users: need adequate access to terminals. |

**Table 3.5 (cont.)** Impairments and disabilities associated with ageing and suggestions for guidelines for accessible ICT design

| Physical capability | Psychological function | Common effects of ageing | Suggestions for guidelines |
|---|---|---|---|
| **Cognitive system** | | | |
| Long-term memory | retrieving memories over considerable periods of time | better for older memories than more recent ones | Use incidents from early life for developing passwords as an alternative to PIN numbers. |
| Information processing speed | speed at which one needs to read, think, make decisions, calculate | slows, particularly after age 75 | Allow more time both for time-critical tasks and general presentation of information. |

CHAPTER FOUR

# On the Validity of Design Guidelines and the Role of Standardisation

Floris van Nes

## 4.1 INTRODUCTION

This chapter relates to the process of designing products or applications at large. The 'design of HCI', with its guidelines, is a part of that process. It may or may not be successful in terms of enabling the user, any user, of the product or application to use it in a satisfactory way. The fact that the usability of products involving HCI is often subject to criticism, in the popular press as well as the professional literature (Isaacs *et al.*, 1996), appears to be a good reason for reviewing critically how the design process of these products is structured, including all trade-offs considered and decisions made.

Here, however, we will focus on the role played by HCI, or rather user–system interaction guidelines. Why do they exist at all, what are possible causes for limits to their effectiveness, and can these limits be shifted or even removed?

After these general considerations the spotlight is on two classes of users who are often said to have special needs as to usability: older people and disabled users. The special needs will lead to special guidelines for interaction. Again, possible limits to their effectiveness, in terms of being applied and leading to more usable products, are considered, as well as shifting or removing such limits.

The main thread of this chapter is that user–system interaction design guidelines for older people or disabled users can only make a real impact by their being standardised, preferably internationally, followed by legislation based on these standards.

## 4.2 ON GENERIC USER–SYSTEM INTERACTION GUIDELINES FOR DESIGN

When writing about 'design of human–computer interaction' here, 'computer' should be understood in the very widest sense; the interaction with rather complex consumer products is meant as well, since such products nowadays very often have a built-in or 'embedded' microcomputer. This chapter will, therefore, use the term 'user–system interaction', abbreviated to USI, instead of 'human–computer interaction'.

'System' here stands for the combination of hardware and software that is dealt with by the user, in order to fulfil some purpose. To design such a system is a complex affair. We may well wonder whether it is realistic to expect one designer to have knowledge about all facets of a product he or she is designing. This would include, for instance, all the physical principles behind the product's functioning, its materials and appearance before and after a period of intensive use and finally the way in which it should be, can be and actually is used. That is a tall order for a single designer; hence the existence of design teams, whose members all have their own main area of expertise. One such area could, or preferably should, be user–system interaction in view of its importance. So, in the ideal situation HCI, or USI, is judged as relevant enough to include a professional in that field in the design team. Such individuals in many cases will not need written USI design guidelines because of their experience—and if they deem it necessary to consult such guidelines, they know their inherent shortcomings. One of those is that, of necessity, a set of guidelines cannot take into account the context in which the product may be used. Another one is the existence of conflicting guidelines, either within one set or in two or more different sets of guidelines. USI professionals will be able to judge the validity of guidelines in specific contexts and resolve conflicts between guidelines.

However, the situation is different in the all-too-common case when there is no real USI professional member of the design team. It should be hoped, and would seem logical, that guidelines are more in demand in such circumstances, where their substance is basically not a part of the designer's (previous) knowledge. Therefore, the designer cannot judge the extent to which the information in the guidelines holds, and he or she is in principle helpless when encountering a set of conflicting guidelines. Often, because of time pressures, the designer does not even want to be bothered with carefully considering contexts of use or such a conflict. This situation leads to a strong demand for "cookbook-style HCI design guidelines" (Preece *et al.*, 1994)—while it is generally much better to present high-level guidelines that force the designer to consider the context of use in his or her design case and, accordingly, distil the proper design decisions to be made.

To summarise this section: it is stated that, ideally, user–system interaction expertise is primarily available in the design process by including a professional in that field in the design team who in an informal way may consult USI design guidelines. If the design team does not include a USI professional, it—hopefully—realises the necessity of at least employing USI design guidelines. These should, however, be applied with care. The best USI guidelines are of a high level and provide directing principles (Preece *et al.*, 1994). In other words, they are true *guide lines* for the user–system interaction work to be undertaken in the design process.

## 4.3 OBSTACLES FOR APPLYING USER–SYSTEM INTERACTION GUIDELINES IN DESIGN

There are several obstacles to really applying the many available USI or HCI guidelines in the design process:

1. The sheer number of guideline collections in this field makes one suspicious about their validity—would not one or two be enough? Anyway, the probability of conflicting statements simply increases with an increasing number of guidelines.

2. The experience of the author of more than 25 years in industry has taught him that in practice there are many forces that interfere with the application of guidelines. Apart from the obvious ones such as lack of time or interest, or plain ignorance about the existence of the guidelines, there are more subtle counteracting forces such as other design rules springing from sources such as certain views on aesthetics, or 'design fashion'. The latter is an understandable, albeit regrettable, consequence of the short product cycle times in many industries. For example, a carefully chosen, legible combination of symbol and background luminance and colour, such as black text on a very light grey, may be for reasons of novelty replaced by a so-called 'military look' primarily characterised by very dark grey housings—resulting in a very low legibility of the text, if it remains black.

3. Designers are not acquainted with human factors and data in general, and this is coupled with what may be called a naive egocentricity. Designers often do not seem to be aware that other users of products they designed differ from themselves, maybe in bodily dimensions or in age. The first can, for example, lead to cramped seating spaces in civilian or military vehicles; the latter can lead to the very small font that is used for the names of authors of papers in even some HCI conference programmes! The designers of such a printed programme, probably young and therefore still in full possession of their accommodative eye lens powers, simply may not realise that older readers have difficulties with such small print—even when they are wearing reading glasses. Let it be noted in passing that HCI congresses are, strange though it may seem, a very good hunting ground for examples of poor presentation legibility, notwithstanding their justified high-tech reputation. For instance, replacing slide-projection by projection of electronically generated images often leads to a deterioration of image quality, such as poor resolution, jitter, and poor colour registration.

## 4.4 HOW TO OBTAIN APPLICATION OF USER–SYSTEM INTERACTION DESIGN GUIDELINES—A CASE STUDY

The actual application of USI guidelines by an organisation involves a number of issues. They are described in the following study.

- **Perceived validity of guidelines by different parties**
  Obviously, those who formulate and collect guidelines are convinced of their validity. However, this need not be at all evident to others; the guidelines may in fact seem superfluous to designers and product managers: "Who on earth would not be able to perform these easy control tasks?" Of course this reasoning is incorrect, but experience shows it is hard to overcome. The advocates of user–system interaction, therefore, are wise to direct their plea to higher managerial levels.

- **Organisational structure of product design**
  In order to obtain an embedding of the application of USI guidelines in a manufacturing organisation, its higher management needs to be convinced of the validity of the guidelines as well as the necessity indeed to apply them in

the design process. This was the case with the Audio Business Group of Philips Electronics in The Netherlands. To counteract the sometimes confusing variability in features and controls of today's consumer electronics products, this group took the initiative to promote consistency as well as simplicity of use in all of its products. Adding consistency of use over a product range to ease of use was regarded as a contribution to a clearer product identity for the customer—and therefore also appealing for product managers.

- **Establishment of a proper set of guidelines**
  The first step for this purpose was the specification, through research, of a set of user requirements for the use and control of the family of audio products concerned: CD players, audio-tape recorders and players, tuners, amplifiers, etc. The user interface specification is an essential part of the user requirements specification. A set of guidelines was developed for the user interface specifications for all audio products made by the Audio Business Group (de Vet, 1993) and made available to all its interaction designers and software developers as a booklet. In practice the developers appreciated the booklet more than the designers, for whom it was made primarily. The reason for this was that the software developers regarded it as a clear reference source for details that otherwise are often not described, whereas interaction designers tend to look at the standardised specifications from a set of guidelines as threatening their creativity and hampering innovation. Doubtless these differing appreciations of guidelines mirror the different disciplines represented by interaction designers and software developers.

- **Facilitation of the application of guidelines by designers**
  However, the actual application of guidelines by designers and product managers when developing a new product is facilitated considerably if the design process itself with its concomitant chores is made easier by support from a software tool. Therefore, after the establishment of the interface guidelines these were further developed through computer simulation of user interfaces for audio products. To stimulate and encourage the use of the guidelines, their electronic form was incorporated in an iterative design process (van Nes, 1994; Bösser *et al.*, 1995). The resulting library of user interface software components with its architecture is now available on line on the corporate intranet. This facilitates both consulting the library and demonstrating dynamic simulations of user interface elements and their interaction.

- **A bonus from facilitation: early usability evaluation**
  An additional advantage of the electronification of the guidelines was the possibility to evaluate the resulting user interfaces with usability tests in a very early phase of development. So, a standard usability test procedure was set up. Such standardised tests in fact serve two aims: they permit checks of compliance with the guidelines throughout the product development process, and allow benchmarking with similar products from competitors (de Vet, 1996). In this way the guidelines are in fact built into design tools that not only serve the primary design process, but also allow such secondary aims as checking compliance and providing benchmarks, even before product development is finished.

## 4.5 USER–SYSTEM INTERACTION DESIGN GUIDELINES FOR OLDER PEOPLE OR DISABLED USERS

The 'application case' for such guidelines in principle would appear more favourable than for the general ones. In many instances it is obvious that these users are quite different from the, as a rule, young and able designer. The designer, therefore, will not so easily fall into the trap of thinking that since he or she can control the product or system concerned, everybody can. Sometimes this is entirely out of the question: blind users cannot read anything on a visual display, and deaf people cannot hear any auditory message the system may produce. Then, special measures need to be taken that may not be at all obvious and therefore will induce the designer to look for help, for example, on the possible application of other sensory channels.

However, if the differences between a special user group and the average user are less obvious, as is the case with older people versus younger ones, the designer, or even the design team, may be completely unaware of those differences. We have already mentioned the lack of accommodative power of the older eye lens leading to smaller fonts being difficult or impossible to read. A slight slowing down of neural propagation is another effect of advancing age. It leads to a required increase of necessary delay times on visual or auditory prompts from the system. If this slowing down is not taken into account, the older user of, for instance, an automatic teller machine (ATM) may be left out of cash just because he or she cannot keep up with the pace of the machine's interaction dialogue.

Yet, it seems more likely that designers will follow specific guidelines for special user groups than generic guidelines for all users. These special user groups are smaller than the population at large and can be easily imagined as having special needs, by their disability or being over a certain age. One may ask, however, whether enough data are available to put into those guidelines or, if not, whether sufficient research effort is directed towards generating such data. The latter issue was the reason for, among others, the TIDE Programme: Technology Initiative for Disabled and Elderly people, from DG XIII of the European Commission. One of the TIDE projects, named USER—an acronym for User Requirements Elaboration in Rehabilitation and Assistive Technology—led to the USER*fit* guidelines, described elsewhere in this volume (Poulson *et al.*, 1996).

A further TIDE project was directed towards enabling blind computer users to keep up with the developments in information technology that started in the beginning of the nineties. The part of this project that was carried out by IPO (Center for User–System Interaction) in Eindhoven, The Netherlands, later developed into a PhD project, culminating in a doctoral dissertation entitled *Visualising Graphical User Interfaces for Blind Users* (Poll, 1996).

Another European initiative was Project 219 of the European Cooperation in the Field of Scientific and Technical Research, COST. Project 219, or COST 219, was called *Future Telecommunication and Teleinformatics Facilities for Disabled People* when it started in 1987. Its successor, COST 219bis, is called *Telecommunications Access for Disabled People and Elderly*. Its activities are also described by Ekberg and Roe in Chapter 13.

From this brief description of TIDE and COST 219bis it may be concluded that there is, at least in principle, neither a lack of interest in trying to identify the necessary data to be put in specific user–system interaction guidelines for older people or disabled users, nor of required research funds for generating these data if they are not yet available. In addition, a project such as USER with its resulting USER*fit* guidelines shows that, again in principle, such data can be arranged so as to form a collection of specific guidelines. Their application by designers, however, meets not only the same obstacles as were mentioned in section 4.3, but some additional ones that may grow into barriers—to be described in the next section.

## 4.6 BARRIERS TO THE APPLICATION OF USER–SYSTEM INTERACTION GUIDELINES FOR OLDER PEOPLE OR DISABLED USERS

The example from the previous section on enabling blind people to use graphical user interfaces (GUIs) is of interest in elucidating the sometimes seemingly diverging interests of people with and without disabilities when a technology develops.

In the time that character-oriented computers were the state of the art, the texts on their visual displays could be converted to synthetic speech or Braille, and thus employment was provided for a number of blind people. However, to benefit sighted users, also making better use of the possibilities of computers, their character-oriented interfaces were generally replaced by graphical ones, GUIs. This caused great problems for partially sighted users, and certainly for blind users, who suddenly saw their jobs endangered, together with everything that is connected with paid work in our society.

Another, slightly older example of a technological development having bitter consequences for people with disabilities is provided by telephony. Its widespread use first meant an excellent opportunity for work for the blind, for example, in private or public telephone exchanges. But the introduction of text displays in telephone instruments, especially in the ISDN (Integrated Services Digital Network) threatened to eliminate this work again (van Nes and Bouma, 1990). On the other hand, one should realise that deaf people benefit from the introduction of text displays on telephones.

The last example from telephony shows how a technological development may first create jobs, but later destroy them again. The first one, from computing, shows how this technological development may make it more difficult than it was to find 'output substitutes' for people with a sensory disability. Of course this is a problem at a different level than that of the 'obstacles' from section 4.3—but the problem is very real. The question is: will an existing guideline, say on replacing graphical user interfaces for those who cannot utilise graphical representations, with their underlying principles for organising data and applications, be searched for and eventually applied? The answer depends primarily on costs—as well as on directives to do so.

As to costs, it is useful to distinguish between the cases where a user interface needs only moderate modifications, for instance increased letter size, and those with more substantial changes, such as replacing a graphical user interface by one employing tactile and/or auditory means. In the first case, there is considerable evidence (van Nes and Bouma, 1990; Abernethy and Lint, 1999) that the modified interfaces are appreciated by all users, that is to say, also by users who do not have the disability concerned. Therefore, any costs incurred in making the modifications may well pay off in terms of increased, widespread user satisfaction and, ultimately, increased sales of the product concerned.

The second case, that with the substantial changes, leads in general to higher costs that moreover have to be made for fewer customers and products. So the manufacturer of those products may shy away from making these changes. This is especially the case if the designer feels uncertain about the adequacy of the adaptations, for instance because he or she does not know whether a given guideline can be trusted in an area that is rather unknown to him or her. This is where standardisation comes into play.

## 4.7 STANDARDS, DIRECTIVES AND LAWS

Standardisation of guidelines increases the trust of designers and manufacturers and makes such guidelines more widely known. This holds for both generic and specific user group guidelines. In particular, the international standardisation procedure—scrutinising, discussing, modifying and eventually agreeing upon the material to be standardised through national votes by experts and the parties concerned from different countries—gives a reasonably good guarantee that the guidelines involved are correct and indeed applicable. So standardisation provides a way to develop a reviewed classification of relevant HCI guidelines. Furthermore, since standards, in particular international ones, are better known than sets of guidelines, they help designers world-wide find relevant guidelines to include the needs of older people and users with disability.

Standardisation increases not only the trust of designers and manufacturers, but also of lawmakers. Through this process the VDU directive (Council of the European Communities, 1990) was compiled by the European Commission from, among others, ISO 9241–3, *Visual Display Requirements*. The VDU directive has the power of law in the member countries of the European Union. Another example of legislation regulating the use of VDTs, visual or video display terminals, from the United States is captured in the 'San Francisco ordinance'. By passing this ordinance on 27 December 1990 San Francisco became the first city in the USA regulating the use of video display terminals in city offices and companies with 15 or more employees (Foreman, 1991).

The legislative process, also in the area of user–system interaction, comes into action when the demand, that is to say public pressure, becomes strong enough. When the lawmakers look for certified information to be put into the legislation concerned, standards, in particular those that have obtained worldwide agreement, are a good source for that information. Such standards can, therefore, be precursors of law.

## 4.8 STANDARDS ON USER–SYSTEM INTERACTION FOR OLDER PEOPLE OR DISABLED USERS

Let us return to telephony or, more generally, telecommunication. This section gives a short overview of relevant telecommunication standards in relation to design guidelines. For a more elaborate treatise of these telecom standards, in relation to market developments, the reader is referred to Chapter 5. The ITU–T, the International Telecommunication Union's Telecommunication Standardisation Sector, is generating so-called recommendations. These have, after the required worldwide approval has been obtained, the status of telecommunication standards. Some of them are directed at the needs of special user groups, for example:

- Recommendation E.135: Human factors aspects of public telecommunication terminals for people with disabilities
- Recommendation E.136: Specification of a tactile identifier for use with telecommunication cards
- Draft Recommendation E.138: Human factors aspects of public telephones to improve their usability for older people (in preparation).

National telecommunication providers are not legally obliged to follow the ITU–T Recommendations. However, they know of their existence and therefore are vulnerable to criticisms and pressure from advocacy groups if the systems they manage do not at least try to accommodate the needs of all users, including those with certain disabilities. Moreover, because of the activities of the advocacy groups or, preferably, an increased moral sense in the political system of a country or groups of countries such as the European Union, a law may be passed to enforce special measures to be taken in order not to exclude user groups from utilising the telephone system, which, as we all know, plays an enormous role in people's communication needs. Such 'special measures' could—just as in the general case described in the previous section—then be easily distilled from the Recommendations being in force, or, in the case of a European country, also from an ETS, a European Technical Standard as issued by ETSI, the European Telecommunications Standards Institute. ETSI has produced a number of ETSs in this area, such as:

- ETS 300 488: Telephony for hearing impaired people. Characteristics of telephone sets that provide additional receiving amplification for the benefit of hearing impaired
- ETS 300 767: Telephone prepayment cards. Tactile identifier (the equivalent of ITU–T's E.136)

and also an ES, ETSI standard:

- ES 201 381: Telecommunications keypads and keyboards. Tactile identifiers.

The last is an example of the results that may be obtained by well-documented criticisms, in this case from the Danish Centre for Technical Aids. Experts from this Centre had done a survey of a large number of mobile telephones and found that many of these did not have a tactile identifier on the central /5/ key. Such a tactile aid for orientation, usually a raised dot, could and can generally be found on the /5/ key of stationary telephones. However, it turned out that there was neither an ITU–T Recommendation nor an ETS or ES specifying the need (for blind people), or desirability (for users who cannot look at or see the keypad while dialling a number) of a tactile identifier for the central key of the keypad. Therefore, ETSI made such a standard: ES 201 381, which also contains guidelines on tactile identifiers for full alphanumeric keyboards. ITU–T will follow by modifying an existing recommendation on keypad layout, E.161. The modification will consist of specifying a proper tactile identifier on or in special cases next to the /5/ key.

Another European standards organisation is CEN, the 'Comité Européen de Normalisation', in English translated to the European Committee for Standardisation. One of its general European standards is:

- EN 1332: Machine-readable cards, related device interfaces, and operations:
  Part 1 Design principles and symbols for the user interface
  Part 2 Dimension and location of tactile identifier for ID cards
  Part 3 Keypads
  Part 4 Coding of user requirements for people with special needs.

The significance of Part 2 of this European standard is that it specifies a particular notch at one position, viz. the right-hand short end of the card, so that blind users always know how to position their cards, either for making a phone call from a public telephone or for obtaining money from a cash machine or ATM—provided of course that the machine readers of those cards also have a standardised position and orientation (see Figure 14.6 in Chapter 14).

In the foregoing section, it was assumed that the standardisation process as such always works to the benefit of older people or disabled users. However, it is also possible to take the opposite view: that standards, of either a formal or a de facto nature, can erect barriers for those special user groups. In response to this view, ETSI organised a European policy workshop on standardisation and disability on behalf of the European Commission, where experts met and discussed "how standardisation processes can be improved so that future standards do not build up new barriers for people with disabilities or older people" (Brandt, 1996).

## 4.9 CONCLUSION

The issues of finding, verifying and effectively applying guidelines for the design of user–system interaction, especially for older people and disabled users, may be analysed. Consequently, finding such guidelines in the literature, or even undertaking appropriate research to gather them, turns out to be less of a problem than applying them in the actual design process. Verification of those guidelines generates trust in product managers—and thus is a step in the direction of having them applied. This verification is probably best reached by efforts to arrive at standardised guidelines, preferably international standards on user–system interaction design for the envisaged target groups. Such standards also form a solid basis for legislature, which ultimately may be necessary to promote the usability of all, or at least most products for all people, based on information and communication technology.

## ACKNOWLEDGEMENT

The author wishes to thank his colleagues Mathilde Bekker and John de Vet for their valuable comments and suggestions on an earlier version of this chapter.

CHAPTER FIVE

# Markets and Regulations

Knut Nordby

## 5.1 INTRODUCTION

During the past 20 years the telecommunications and IT markets have undergone considerable changes that are characterised by *market liberalisation*, *privatisation*, *regulation* and *standardisation*. Gone are the national, public telephone agencies; gone are the stubborn, cavilling and bureaucratic monopolies; and gone is the humble standard telephone that came with your subscription. Enter the private telecom operators and the possibility to choose your service provider; enter the big transnational corporations, fiercely competing to dominate the market. Gone also are punch-cards and batch-processing on mainframe computers. Enter the 'Mac' and the PC and the purpose-designed software applications; and, not least, enter the vast undergrowth of small enterprises, providing a host of new services and the huge variety of equipment now available in any shopping mall.

How do these market changes influence the situation for disabled and older customers? Has this development been for better or worse for these groups? In this chapter[1] I hope to show how the *market*, *competition*, *liberalisation*, *privatisation* and *deregulation* of converging telecommunications and IT (into information and communication technologies or ICTs) changes the situation for older people and disabled people. I will attempt to show what can be done to protect users from any ill-effects of the market by means of such remedies as *guidelines*, *procurement*, *regulations*, *standards* and *legislation*.

## 5.2 THE MARKET

Since the old public telecom monopolies were general service providers, some of them felt that they had obligations to all their subscribers. They would therefore provide facilities for disabled subscribers in the form of telephones with extra amplification for the hard of hearing, telephones with large dials for the visually or motor-impaired, text-telephones for deaf people and free directory enquiries for blind people. Practice varied greatly—while some provided these facilities free of charge for those who needed them, others charged for their use, while others again did not provide anything (Tetzchner and Nordby, 1991).

The IT industry has been very slow in taking the needs of older and disabled users into account because it has had no past as state-owned enterprises with self-

[1] This chapter is based on a presentation at a COST 219 Seminar *Human Aspects of Telecommunications for Disabled and Older People*, 11th June 1999, The University of the Basque Country, Donostia/San Sebastián, Spain.

imposed responsibilities to their customers. Using a computer was the domain of specialists and the rapid spread of personal computers was not envisaged. Thus there was no perceived need for any adaptations for older or disabled users. The introduction of graphical user interfaces and the direct manipulation of icons with pointing devices which was hailed as the greatest innovation in personal computing, was a disaster for blind and visually impaired people. Over-busy Web pages on the World Wide Web are yet another obstacle for visually impaired and older users. Although IT is most proficient for implementing solutions to overcome different disabilities, it is the standard products from a few leading software developers that dominate the market. The big concerns have been slow to recognise the problems for older and disabled users and press ahead for greater profits in a market that does not seem to have any limit. They may have to pay dearly for their hubris if legal actions are brought against them.

Fifteen years ago most European countries had public telecom monopolies, generally of little or no commercial value. Technologies were largely outmoded—twisted-pair copper wires, mechanical switching and predominantly analogue transmission. Operating these telecom monopolies was very labour-intensive and bureaucratic. They were not expected to give their owners any profits and were dependent on large public subsidies. The reason they were kept running was of course that telecommunications are absolutely essential for the functioning of any modern industrialised society.

The massive introduction during the past 15 years of new technologies, such as microprocessors, fibre-optic cables, digital transmission and switching, ATM (Asynchronous Transfer Mode), broadband radio, cellular radio networks, etc., and of new services, such as ISDN (Integrated Services Digital Network), wide-band data transmission, multimedia, Internet, e-mail, etc., has significantly changed this situation. From being regarded as rather backward, non-profit mammoths, the modernised telecom agencies were soon perceived as profitable and commercially interesting enterprises—and they caught the investors' attention. Market forces started to exert pressure on governments and legislative assemblies to liberalise and deregulate the telecommunication market and to privatise the public telecom monopolies, and a wave of liberalisation and privatisation swept Europe and other parts of the world. This was in the halcyon days of 'Thatcherism', and involved not only the telecom monopolies, but also other public utilities such as electricity boards, gas boards, waterworks, railways, etc.

At first, the state continued to own all the shares in the privatised telecom agencies, or at least to keep a controlling majority. The current trend, however, is to go public and to sell down, offering shares both to employees and to investors at large. Market forces soon led to mergers of telecom companies and to the buying up of smaller companies by larger transnational concerns. If the rate of mergers continues, we may soon have only a few 'mega-telecom' operators left in the world, each serving a separate region.

The privatisation of the monopolies and the liberalisation and deregulation of the markets was intended to give customers a real choice of different service providers, and the subsequent open competition was believed to lead to lower prices and improved services. To some extent this has happened; prices have been significantly reduced, there is now a real choice between service providers and new services have been introduced. However, there has been a noticeable decline in service quality (e.g., quality of speech in mobile telephones), and in the services offered to older and disabled customers. Growing larger to control the market is the primary aim of big corporations which are in the market for one reason only—to serve their owners by maximising profits. It seems that the larger a corporation is, the less inclined it is to listen to the needs of older and disabled customers.

These large corporations may soon take on the appearance of monopolies and will dictate their own terms. This may lead to diminishing real competition, so there will be no guarantee any more of lower prices or improved services (compare the position of the leading software concerns in the IT world). Eventually, this may lead to the use of anti-trust and anti-cartel laws to break up the large monopolies, like the huge Bell Telephone Company of USA being broken up into 'Baby-Bells' in 1984.

However, the liberalisation of telecommunications has also opened up the market to a large number of small entrepreneurs with innovative ideas who can now put their services on the network and their equipment into the shops. Usually these entrepreneurs fill niches ignored by the large companies. Sometimes they develop important new service concepts or products, but are then more often than not bought up by the large corporations. Often the small companies come up with well-thought-out products or services for older or disabled customers—products and services that the big operators do not bother to offer if they are not compelled to by law or by the licensing requirements for granting the right to operate telecom services.

Despite this, the situation for older and disabled people has gone from bad to worse. While the old state-owned monopolies sometimes provided at least some services for these groups, the privatised operators, with one or two commendable exceptions, regard such services as the province of social welfare, the medical or education authorities, the insurance companies or even the charities. Provision of special services and equipment for older and disabled customers is regarded as irreconcilable with profit maximisation. However, older and disabled customers do already comprise over one quarter of the private household subscriber market, and the share of older customers is growing very fast.

As general health is constantly improving and people live longer, and as the age of retirement is reduced and pensions increase, the number of these attractive 'Golden Oldies' is increasing. They already constitute an important market segment that has mostly been ignored by the big companies. When industry eventually realises the market potential of older customers, it must start to pay attention to their needs. These stem from age-related impairments of vision, hearing, the haptic (touch and feeling) sense, motor, memory and cognitive functions. If these needs are correctly handled by providing improved services and equipment, this will also be of great help to many permanently or temporarily disabled users, since their needs are mostly the same as those of older people with impairments, thus further increasing the market.

With a few exceptions, industry has not recognised the market potential of products for these consumers. There are examples of products that were thought of as specialist products but that subsequently made it into the marketplace. One good example is the 'talking book'. For many years the societies for blind people provided books read on to audiocassettes as a special service to their members. For a long time it was not realised that this product had any commercial potential outside the community of blind and visually impaired people. Today, the talking book (on CD or cassette) has become a thriving mainstream product that is now available at most bookshops and CD retailers.

There is an important lesson for industry here. Most people will, in specific situations or at some time in their lives, be temporarily impaired, e.g. in darkness, in very noisy environments, in extreme cold, following an accident or during illness. Products that are designed for people with impairments will often be of great benefit to people who are temporarily disabled but who do not consider themselves old or disabled. The wider market potentials of products designed to take care of the needs of older or disabled people have not been well understood by industry but as the example of the talking book shows, industry will have to change its attitude to these products. There are unexploited market sectors in ICT for manufacturers and service providers that take the trouble to identify them (Martin, 1991; Perret, 1995).

We find a very different situation with the Internet and the World Wide Web. The Internet is not 'owned' by anyone and is largely self-regulated. It is also a vehicle for other markets (electronic mail, electronic commerce, electronic banking, etc.) and it is thus in the interest of all parties involved to keep the Internet as an open system to serve its users for various activities, both commercial and non-commercial.

The market will thus function in very different ways for older and disabled people. On the one hand, by not providing special services in an effort to maximise the short-term profits to their owners, industry can make the situation more difficult for older and disabled people but may then run the risk of regulatory intervention or legislative actions. On the other hand, by identifying the special needs and utilising their market potential, it can improve the situation for older and disabled people, providing long-term profits to their owners and avoiding any requests for regulatory or legislative intervention.

## 5.3 REGULATIONS

The past ten years have seen both *de-regulation* and subsequent *re-regulation* of the telecom markets. In Europe, most countries have transferred much of their regulatory powers to the Commission of the European Communities (CEC), which can issue CEC Directives. Still, the European market is less regulated in many respects than the US market, which in Europe always has been regarded as one of the most liberal markets in the world.

When the telecom monopolies in Europe were privatised, national regulatory bodies were set up in each country to deal with the new situation. Regulations are, of course, essential for co-ordinating services, networks and the interworking of different systems, and, not least, to regulate the competition between independent operators. However, operators and manufacturers demanded unhindered trading across the whole open European market and any local national regulations that obstructed free trade were deemed unacceptable by the telecom industry. The CEC has therefore issued CEC Directives to regulate the European telecom market. These Directives often take into account standards developed by the European Telecommunications Standards Institute (ETSI), which had been set up in 1988 to create standards for the open European market. However, the telecom industry found even the liberal Directives too restrictive on trade and too slow in the making and demanded less national and international control. The latest CEC Directives allow a wide degree of self-reporting. Thus, the telecom industry may now decide itself whether it complies with standards and regulations or not.

This is completely unacceptable to consumers. All experience so far has shown that such freedom invariably will be abused by industry at the expense of consumers in general and of older or disabled consumers in particular. If a telecom service or product does not function or follow specifications as advertised, the customers need protection from a fully independent agency that has the formal mandate and authority to deal with the telecom operators and manufacturers. If this development continues unchecked it may, in the future, be impossible to obtain compensation for malpractice from the telecom industry. Even now it can take several years and cost tens of thousands of Euros to obtain a court ruling on a bad product, costing maybe less than 100 Euros and with a product-life of only ten to twelve months. With competing companies constantly merging or being acquired by larger concerns, the original service provider or producer may not even exist to recompense the customer by the time a court ruling is obtained. There is therefore an urgent need to reregulate the telecom market to protect consumers' rights. Consumers need a watchdog that is authorised to monitor the telecom industry and to enforce the licensing requirements for granting the right to operate telecom services in that country and that has the judicial power to bring sanctions against those companies that do not comply with regulations and standards.

## 5.4 PUBLIC PROCUREMENT

Since industry is so concerned about profits, the simplest way to make companies comply with guidelines and standards is to hit them where it hurts most—namely economically. This can best be achieved by referring to relevant guidelines and standards into all *public procurement* calls for tenders for equipment and services. These calls for tenders should be based on the accepted guidelines and formal standards and should then treat all contenders fairly—those that comply with the requirement of the tender are allowed to compete on equal terms, those that do not are excluded from competing for the tender. The guiding principle should be that if the intended users cannot use a particular product it should not be bought. In fact, large amounts of equipment—often unsuitable for older or impaired users—are bought by public organisations, wasting taxpayers' money.

## 5.5 GUIDELINES

Guidelines are sets of guiding rules for *accepted good practice* that the industry may follow at its own discretion. They often build on standards and like these can come from formal or self-appointed groups. When developing usability guidelines it is essential to take into account how the market functions. Industry does not want to be told what to do and generally resents any regulatory actions or mandatory standards—most companies hope that their products become market leaders and the *industry* (*de facto*) *standard*. However, in some cases the industry does accept standards, e.g. to ensure interoperability or interchangeability of parts. However, the industry may follow non-mandatory guidelines so long as it believes there is some real gain. Thus, the industry can sometimes have a positive attitude to guidelines, not least because of their optional and voluntary nature.

## 5.6 STANDARDS

There are basically two kinds of standards: *formal* (*de jure*) *standards* and *industry* (*de facto*) *standards*. Industry standards arise when a product establishes itself as the market-leader and is recognised by the rest of the industry as 'the standard'. Formal standards, on the other hand, are drafted and adopted by formally appointed standards organisations. Most formal standards are based on *consensus* of the members of the standards organisations, but European Norms (EN) are voted over by official national bodies. Most formal standards are *voluntary*, which means that they only need to be followed at the discretion of the manufacturer or service provider, unless the standard has become part of a law or a CEC Directive (Ekberg *et al.*, 1991; Martin *et al.*, 1995).

The first telecom standards were national. However, there was early on a need for international rules to make telegraphy between different countries practicable, and on 17$^{th}$ May 1865 the International Telegraph Convention was signed by 20 countries. Later that year the International Telegraph Union (ITU) (later changed to International Telecommunication Union, also known for many years as CCITT) was founded to deal with subsequent amendments of the initial Convention and to adopt new telecom recommendations as required. Regional telecom standards bodies have been established in many parts of the world to create standards for their respective regions, e.g. ETSI (Europe), ANSI (USA), ADD (Australia), JDD (Japan) and KDD (South Korea). However, there is an increasing need for global telecom standards and ITU–T (International Telecommunication Union's Telecommunication Standardisation Sector) may soon come to play a much more important role than previously envisaged (Nordby, 1991).

Contrary to industry, the standards organisations have obligations to the users, especially to older and disabled users. A number of standards have been created to make telecommunication services and equipment accessible to these user groups. Since most standards are non-mandatory, industry does not have to follow them. In the rapidly changing telecom world where market share and maximisation of profits decide what is 'best', industry has shown little interest in participating in creating the required standards for these user groups.

As mentioned above, standards are based on consensus in standards bodies. This was much easier to obtain in the days of the state-run telecom agencies than it is today, because they had no conflicting commercial interests. In contrast to the situation in the telecom world, the Internet world relies on *de facto* standards. Internet and WWW 'standards' are developed by bodies with no formal mandate, e.g. IETF (Internet Engineering Task Force) and similar *ad hoc* groups. These groups convene as they find appropriate and decide by consensus or simple majority how things are to be done, with no allegiance to any formal standards organisations or regulatory bodies. Sometimes *de facto* standards are taken up by formal standardisation bodies and turned into *de jure* standards.

With convergence of traditional telecom technologies and IT (e.g., telephony over the Internet, data over mobile phones, etc.), a new situation with more *de facto* and fewer *de jure* standards may soon be the norm. This will most certainly have a great effect on the traditional telecom standards bodies and completely change the role of the official standardisation organisations, unless some super-national body can take charge and produce the necessary regulating legislation. Although ITU–T is a United Nations organisation, it has not been able so far to deal decisively with convergence of telecommunications and IT and the rapidly growing Internet. There is a need for global standards and for an internationally recognised official body to regulate all forms of telecommunications, including the Internet, for the benefit of society, industry and users.

## 5.7 LEGISLATION

Until now, the privatised telecom industry has shown little or no interest in dealing with the requirements of older and disabled customers by providing the necessary facilities. Even though many standards for older and disabled people have been adopted, the standards need not be followed by industry unless they have been made part of laws and directives or are specified in calls for tenders. Operators and service providers in the USA have to provide special facilities and equipment for older and disabled people because the Americans with Disabilities Act (ADA) of 1990 and the Telecommunications Act of 1996 require them to do so. However, these very same companies will not provide the same special facilities and services in European or other countries where no such legislation exists. However, even large international concerns have to abide by the laws of the countries in which they operate, laws that either grant special rights to older and disabled people or regulate the licensing requirements for granting the right to operate telecom services in that country.

If older and disabled people are ever to have ICT facilities and services that are adapted to their needs, the only certain way will be through legislation—national or international. Work has already begun to create European legislation similar to the ADA and Telecommunications Act in the USA. Such legislation may take some time to be adopted. This is largely due to the slowness of the European legislative process based on consensus in several countries, and, not least, because of inherent differences between the US and European judicial systems (Cobut *et al.*, 1991; Kemppainen *et al.*, 1995).

## 5.8 CONCLUSIONS

Meanwhile, there are efficient means to achieve facilities for older and disabled people, viz. *licensing requirements, procurement specifications* and *guidelines*. Any country can adopt its own *licensing requirements* for granting the right to operate telecom services in that country, as long as they do not interfere with open competition. Licensing requirements should contain obligations to provide specified services and facilities for older and disabled people. Such requirements should not affect competition between telecom operators, since the same rules will apply equally to all the actors in the market. The requirements should be based on international standards and guidelines.

An alternative means to achieve better services and facilities for older and disabled people is by procurement specifications, i.e. requesting that all applicable standards and guidelines are followed in public calls for tenders. The risk of lost sales may, in fact, turn out to be the most effective instrument to motivate the ICT industry to comply with existing standards and guidelines. Such actions would apply equally to all the contenders in the marketplace.

Standards and legislation may, of course, also function as guidelines for industry. It is, therefore, most important to develop well-accepted, sound, clear and unambiguous technical specifications and standards that are easy to understand and simple to follow by the designers and that are accepted as relevant by industry. Guidelines, however, have no strong formal status, and this is both their strength and weakness. On the one hand, industry *may* follow them as long as they are not mandatory, which is good for users; on the other hand, the lack of any formal status means that we have no means to force industry to follow the guidelines, however relevant they may be.

The ICT market and our means to regulate it is a very complex system, and I do not presume to have more than scratched the surface of the subject. Awareness of possible pitfalls is essential when trying to impose guidelines on industry. It is not difficult to create good guidelines and standards for inclusive design to make ICT more accessible to older and disabled users—the real challenge is how to encourage industry to comply with them. Ultimately, as shown by the American ADA and similar laws, I see no alternative to legislation. Such legislation has been discussed in COST 219 and a way has to be found to urge the CEC to act decisively in the matter—this legislation is already long overdue in Europe.

# Part 3

# Tools for Accessing and Using Guidelines

CHAPTER SIX

# Managing Accessibility Guidelines during User Interface Design

Constantine Stephanidis and Demosthenes Akoumianakis

## 6.1 INTRODUCTION

Guidelines constitute a popular means of providing input for human factors design in the development of interactive computer-based products and services. In the mainstream human computer interaction (HCI) field, they have been in use for a number of years and have contributed substantially to the development of a human factors *culture* within commercial organisations and institutions. In the field of assistive technology, guidelines provide a consolidated body of knowledge reflecting the design of suitable computer access systems. Today, much of the accessibility work is based on general or context-specific guidelines developed from experience, practice and, sometimes, experimentation.

The term *guideline*, in the present context, entails all forms of abstract or concrete recommendations that may be used to design user interfaces accessible by diverse user groups, including people with disabilities. Such guidelines may be expressed in relation to *universal design principles* (Story, 1998), as *general recommendations* (Rahman and Sprigle, 1997; International Organisation for Standardisation, 2000; Thorén, in Chapter 9) or as platform-specific design rules.[1]

In this chapter, we present a methodology and a supporting software platform that augment the way in which accessibility guidelines are being used and facilitate a more active and direct account of the embodied human factors principles. The chapter is structured as follows. The next section (section 6.2) discusses some of the reasons which impede the use of accessibility guidelines and motivates the adoption of information technology as a natural medium to facilitate the growth and flow of accessibility knowledge. Then, in section 6.3 we propose a methodological framework for managing accessibility guidelines. Section 6.4 describes practical experience with a tool environment which was developed to support the methodology. The chapter concludes with a discussion and summary of the results presented.

## 6.2 SHORTCOMINGS OF GUIDELINE REFERENCE MANUALS

In the recent past, guidelines have been used to develop interactive computer-based systems accessible by different users, including people with disabilities. Despite their sound human factors input, accessibility guidelines are frequently ignored by

---

[1] Examples available at www.ataccess.org/design.html and www.w3.org/WAI/References/

designers. It is argued that this is due to a number of problems which limit the adoption of HCI guidelines in general. In this section, an attempt is made briefly to summarise some of the main impediments to adopting guidelines, and in particular accessibility guidelines, during the user interface design process, as well as to identify means of potential improvement.

First of all, accessibility guidelines, in the majority of cases, are expressed as general recommendations independent of context. This raises a compelling need for interaction- and collaboration-intensive interpretation of the relevant statements. However, any interpretation effort is bound by the capability, experience and breadth of knowledge of the designer (or the specialist involved) regarding alternative access solutions, technical characteristics, etc.

Secondly, accessibility guidelines are not experimentally validated. This relates to a more general problem which arises from the fact that the currently available experimental work in the field of accessible HCI design is generally considered as rather limited and does not cover the broad range of alternative solutions. Additionally, due to radical changes that occur in the mainstream information technology industry, some of the past experimental results rapidly become invalid or out of context.

Thirdly, accessibility guidelines are not easily communicated to development teams. Recommendations derived from guidelines may not always be comprehensible or appropriated by the development team. This is not only due to the typically demanding task of implementing these recommendations, but also due to (a) the doubts that are frequently expressed regarding the validity of a particular recommendation in a given design case; and (b) the vocabulary or language used, which is not always comprehensible and which necessitates additional training before the development team can effectively and efficiently use a guideline manual and implement the relevant recommendations.

Fourthly, accessibility guidelines are difficult to implement as they require substantial programming-intensive efforts. As the industry lacks high-level tools for developing accessible user interfaces, the implementation task required to account for accessibility recommendations is complex. It typically requires such an intensive programming effort which, in general, cannot be reused across design cases. As a result, the cost of developing and maintaining the interface is high and it increases dramatically with the number of different target user groups.

Finally, design recommendations derived through different sets of guidelines are frequently conflicting (sometimes such conflicts may be encountered even within the same set of guidelines). The implication is that a particular guideline may invalidate the recommendations associated with another guideline, thus introducing design inconsistencies. At the same time, guideline documents or reference manuals offer no natural way of resolving ambiguities, which typically leads to arbitrary decisions by the design team.

Due to the above reasons, accessibility guidelines are frequently ignored or misused by designers, especially when guidelines cannot be intuitively translated into unambiguous design specifications with explicit content and scope, or when the guidelines reference manual becomes voluminous.

It should be noted that the above critique does not intend to question the human factors content of accessibility guidelines and, consequently, it does not question the fact that such guidelines provide a collective memory of accessible

HCI design wisdom. Instead, our claim is that guidelines should be delivered in a medium that facilitates their *incremental evolution* as new experience is gained, and *flow* within an organisation to facilitate persistent use and knowledge construction. To this effect, guidelines should be delivered through appropriate and argumentative media that facilitate the growth and flow of accessibility wisdom. Information technology is used pervasively in organisations and, thus, qualifies as a natural medium for both these targets.

Consequently, the normative perspective adopted in this chapter is that software tools can substantially ease the task of managing accessibility guidelines, provided that the latter are developed in such a way as to become an integral part of the organisational culture.

## 6.3 MANAGING ACCESSIBILITY GUIDELINES: METHOD AND TOOLS

### 6.3.1 Some definitions

Before presenting the details of the approach used to arrive at experience-based accessibility recommendations, it is perhaps useful to introduce some terms of reference to facilitate a better understanding of the context of the work elaborated in this chapter. To this end, the term *guideline* is used to refer to a general rule of thumb which applies in a particular context for a particular class (or range) of computer artefacts (i.e., input/output devices, windows, text, etc.). Usually, a guideline is a consolidated statement of existing design wisdom and there is sufficient evidence to support its applicability.

A *design heuristic* is a rule based on experience and may not have been subject to empirical validation, as its applicability may be dependent on the context of use (i.e., user profile, task requirements, usage pattern, etc.).

Finally, *recommendations* reflect detailed design decisions which are usually related to implemented counterparts of design concepts. It is important to mention that recommendations represent unambiguous statements about either the physical (lexical) or the syntactic level of an interface and, once derived, they can be directly implemented by a user interface developer.

The emphasis upon the physical and syntactic levels of interaction is justified by the fact that accessibility typically affects both the choice of physical interaction resources (i.e., input/output devices, object classes and interaction techniques) as well as alternative dialogue organisation.[2]

### 6.3.2 Method overview

This section describes a process-oriented methodology for managing accessibility guidelines during user interface design. A schematic description of this methodology is depicted in Figure 6.1. As shown, the methodology comprises three

[2] It should be noted that in certain cases accessibility may entail alternative interaction semantics (e.g., alternative suitable metaphors), but this is arguably the least studied dimension.

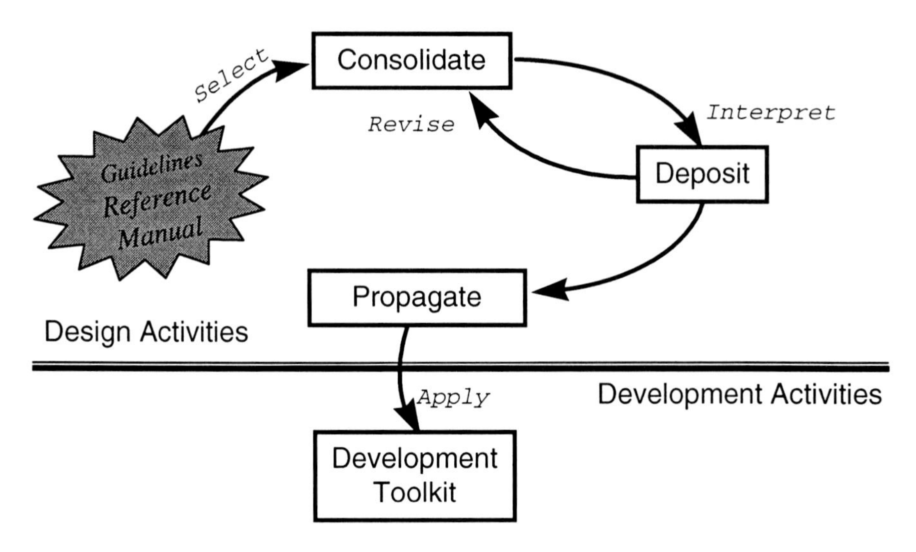


**Figure 6.1** Methodology overview

distinctive design-oriented phases, namely *consolidation*, *deposition* and *propagation*, as well as a bridge with the implementation phase.

Though the methodology does deviate from previous guideline management models, it has an explicit computational flavour. This is to say that computational tools, which are intended to facilitate the growth and flow of accessibility design knowledge, support all three methodological phases. These tools are strongly linked to the paradigm of knowledge-based systems as they embody various knowledge representation schemes and inference mechanisms to facilitate the intended objectives. In what follows, we briefly describe the constituents of each of the phases depicted in Figure 6.1, while in the next section we elaborate on each of them by means of a reference case study.

### *6.3.2.1 Consolidation: selection and interpretation*

The consolidation phase involves two distinctive stages, namely selection and interpretation. During the first stage, the relevant guidelines are selected and compiled to constitute the guidelines reference manual. The second stage entails the interpretation of the relevant guidelines, based on known or domain-specific heuristics. Such heuristics subsequently give rise to platform-specific recommendations. It is important to mention that, when interpreting the guidelines, designers should seek advice from all parties involved, including end-users and developers. Moreover, the selection of suitable heuristics to guide the interpretation process should meet the wider agreement of these parties. This will help the team to establish a common vocabulary which is critical for the subsequent phases of deposition and propagation.

#### *6.3.2.2 Deposition*

Deposition refers to the phase where the recommendations for a particular artefact are encoded into design repositories, using suitable tools and representation mechanisms. This is a technical stage which designers frequently find difficult to comprehend, especially if it entails platform-specific attributes. The challenge during this phase is to decide which interaction aspects are influenced by the guidelines and how such relationships can be captured into reusable design repositories. In currently prevailing practice, this phase is loosely (if at all) carried out. It is, therefore, important that it is supported through appropriate design-oriented tools which designers can work with.

#### *6.3.2.3 Propagation*

The final phase is that of propagation of the applicable design recommendations into user interface development so that they can be accounted for by the implemented user interface. As in the case of depositing recommendations, propagation relies heavily upon the availability of software tools to undertake the task. The tools we have been developing for this purpose are described in Akoumianakis and Stephanidis, 1999. Typically, propagating recommendations comprises two distinctive stages. In the first stage, recommendations are extracted and compiled into an internal representation. The second stage involves the translation of this internal representation into a format that is acceptable to the target toolkit with which the user interface is to be implemented. As a result, the software tools that are to be employed for this phase need to possess knowledge as to what interaction elements are supported by the target toolkit, as well as the detailed format to which the generated recommendations should comply.

## 6.4 A CASE STUDY

To illustrate the practicalities of each phase of the proposed method, this section reflects upon a recent design case study. The example is extracted from a more general design case conducted in the context of a collaborative research and development effort (see Acknowledgements). The objective of the case study was to build accessibility features into a windowing environment so as to facilitate switch-based access[3] by motor-impaired users. In this section, we concentrate on a small subset of the case study which was concerned with the design of a virtual keyboard[4] aiming to allow the target users to carry out interactive tasks, such as text editing, window management, etc. In what follows, we provide a brief insight into the rationale of developing such accessibility features, as well as the use of the design tools as knowledge-construction instruments.

---

[3] This is an interaction method which involves the use of binary switches as the primary input device.

[4] A virtual keyboard is a visual representation of a keyboard and allows users to scan and select elements within a selection set so as to carry out specific tasks. The virtual keyboard is a familiar interaction technique for motor-impaired users.

### 6.4.1 Scenario building: envisioning and unfolding design options

The problem at hand was to provide the user interface developers with appropriate tools (i.e., libraries of interaction elements) to build interactive applications that could be accessed by diverse user groups, including people with motor impairments. For the purposes of this chapter, we assume that the target user population (i.e., motor-impaired users) can access the computer through movements of the right hand. The set of reliable contact sites includes the fist of the hand, while the control movement that can be exercised is gross temporal, as opposed to fine spatial, control. Additionally, it is assumed that users are capable of reproducing the movements on demand and in timed patterns, so as to initiate and sustain interaction using a switch-based input device.

For such a user, interaction with a graphical user interface (GUI) entails the consideration of a broad range of design issues. In this chapter, we intend to address only a few of them for the purposes of illustration and completeness of the arguments. Assistive technology specialists were members of the design team. One of the first considerations addressed was the issue of how to provide access for such users to a windowing graphical environment in a manner which would not be obtrusive for an able-bodied user. This question is depicted in the Questions–Options–Criteria (QOC) map (MacLean *et al.*, 1991) of Figure 6.2.

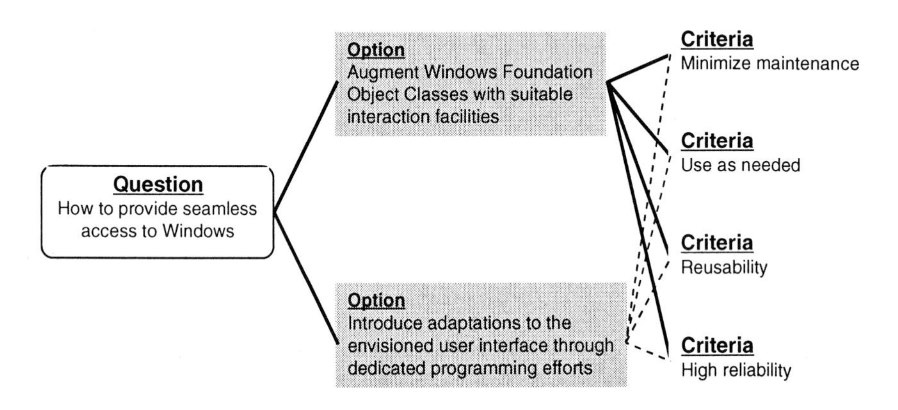


**Figure 6.2** QOC diagram for assessing alternatives for building access to a windowing environment

Two alternatives were identified. The first was to augment the object foundation classes with suitable interaction facilities to allow access. This option was not related to any particular application domain or envisioned user interface, as the new (i.e., augmented) interaction techniques would be embedded into the windowing environment through sub-classing of the existing foundation classes. The alternative was to investigate the envisioned application domain and the user interface so as to design adaptations through a programming intensive effort. The criteria which were raised by design participants and which were also documented in the technical deliverables of the project were those depicted in Figure 6.2. As shown, the first option is preferred across all different criteria and was the strategy that was finally selected.

### 6.4.2 Consolidation

Having committed to the above strategy, we then tried to select and interpret relevant accessibility guidelines. An example of a guideline for user interface design regarding moderately motor-disabled users is taken from Smith (1996):

> G: "Provide a method for carrying out mouse, or other pointing device functions with the keyboard, or a keyboard emulator"

According to this method, such a guideline may give rise to several design heuristics based on the interpretation of the guideline in a specific application domain or context of use. Moreover, these heuristics may, or may not, be related to the same interface attribute and may, or may not, be of equal voting power. In cases where, for the same interface attribute, more than one design heuristic is applicable, then the designer typically needs to aggregate the competing alternatives to arrive at a plausible, maximally preferred solution. For example, the guideline mentioned above may give rise to design heuristics at the level of the overall environment or for specific user tasks. At the level of the overall environment the design heuristic may entail that:

> $H_1$: "Switch access to Windows should be allowed for users with physical/motor impairments"

On the other hand, at the level of the user tasks, a design heuristic may entail that:

> $H_2$: "Text editing facilities should be provided via a virtual keyboard accessible through a switch-based interface"

or, that:

> $H_3$: "Window management facilities should be made accessible through explicit function activation"

For the purposes of this chapter, we briefly elaborate the case of window management. In windowing environments, window management refers to tasks such as minimise/maximise/destroy windows, as well as moving and resizing windows. All these functions are carried out through mouse-based operations upon top-level containers.

A typical top-level application window is depicted in Figure 6.3. As shown, access to such top-level containers requires a mechanism to access the window management facilities (such as the minimise/maximise functions, etc.), as well as access to window navigation facilities (i.e., navigating within a window, moving from one window to another, etc.). Figure 6.4 depicts a QOC map for the issue of providing access to window management facilities, typically offered by a top-level container. As shown, two alternatives were considered.

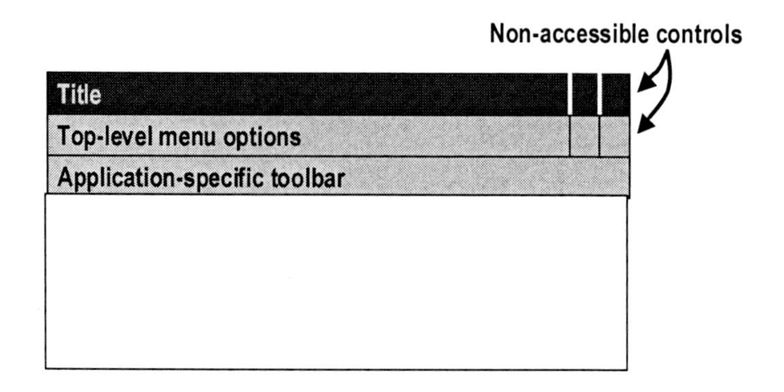


**Figure 6.3** A typical top-level window

The first was to isolate the small window management icons on the top right-hand corner of top-level windows and allow their scanning when focus is on the container object. Though this would require no additional design and its implementation would simply build upon the sub-classing technique, it turned out to be non-feasible for two main reasons. Firstly, early versions of such windowing environments offered no options to access these facilities programmatically.[5] Secondly, even if this were possible, there is still an additional issue of how to re-size windows—something that is not implemented through explicit function activation. In other words, the first option, even if it were feasible, does not provide a complete and consistent solution to the problem of access to window management facilities. Thus, the second alternative was selected.

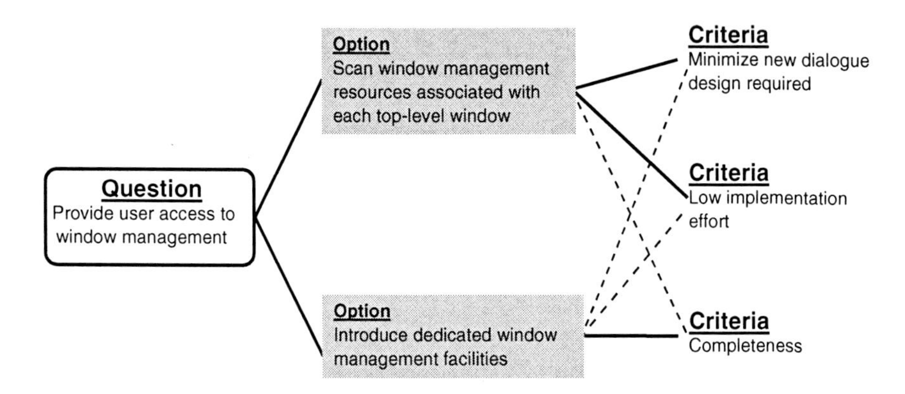


**Figure 6.4** QOC diagram assessing the provision of user access to window management

[5] It should be noted that this deficiency was subsequently removed from most of the popular graphical toolkits.

From the above, it follows that heuristics may imply specific design artefacts which should be developed through a user-involved iterative process (Bannon, 1991) to reflect the requirements of the intended target user groups. For instance, $H_3$ implies the design of dedicated navigation facilities, such as toolbars for certain window management functions. Figure 6.5 illustrates the results of several iterations regarding these toolbars.

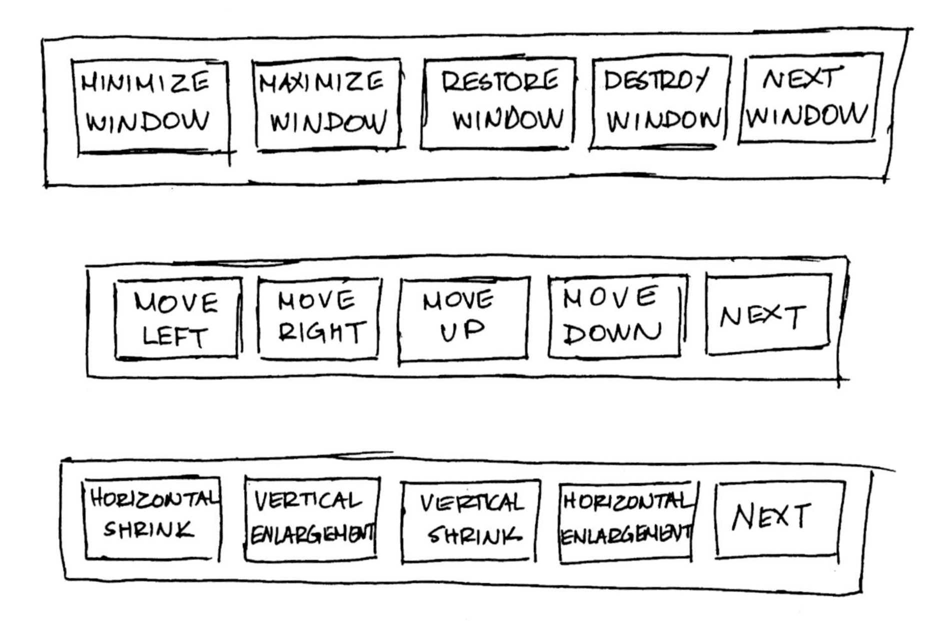


**Figure 6.5** A paper mock-up of the toolbars for window management

### 6.4.3 Deposition of recommendations

Recommendations are typically bound to specific attributes of interaction (either at the physical level or at the syntactic level). For example, recommendations relevant to the physical level of interaction may account for the 'look and feel' of the toolbars designed to facilitate window management, the choice of object classes and specific attributes of such object classes (e.g., the access policy for container objects, the topology of groups of items). Recommendations relevant to the syntactic level of interaction may determine aspects such as the dialogue command order that is to be used to accomplish specific tasks (e.g., Function–Object versus Object–Function syntax), the function activation modes (e.g., explicit versus implicit activation), etc.

Recommendations are encoded in a knowledge representation scheme and stored in a reusable design repository. Though a full description of the properties of this scheme is beyond the scope of this chapter, we provide a brief illustration of the basic constructs. Table 6.1 depicts the representation of the resizing toolbar presented earlier.

An important property of this specification is that it occurs at a level of abstraction that allows for reuse of the same artefact across different design cases. In other words, the details of the toolbar (i.e., layout, size and type of selection set, location of the toolbar, etc.) are extracted from the semantics of the specification which concentrate on the details of the interaction of the user with the toolbar.[6] Thus, what is being represented are additional properties of the artefact that explain how it evolved and how it is going to be used by the target user group.

In Table 6.1, the resizing toolbar is specified as a collection of one container (line 2) and button object class (line 23). Object classes are considered as abstract components which can be mapped to concrete interaction objects in alternative toolkits. Examples of abstract object classes may include generalised containers, buttons, text-entry fields, etc. Moreover, object classes, such as containers and buttons have attributes, which are represented by the predicate `attribute` which is defined as follows:

attribute(symbol,AttrDom), where

AttrDom is α(symbol, string, listOfParameters)

Thus, each object category is modelled as a collection of attributes (in the specification of Table 6.1, lines 3–22 declare attributes for the container), while each attribute may be associated with specific parameters which define exactly how the attribute is supported by various user interface development systems (lines 7–20 declare the parameters of scanning in the container). Attributes are typically classified into three different categories (see also Akoumianakis and Stephanidis, 1997b), namely general (lines 3–6 and 21 in Table 6.1), presentation (lines 24–25 in Table 6.1) and behavioural attributes. General attributes are possessed by all interaction object classes and include non-trivial assignments such as input/output device, interaction techniques, feedback and, for container object classes, navigation, access, topology and access policies. Presentation attributes detail the look of the interaction object class. Finally, behavioural attributes are used to capture behavioural aspects of the interaction.

The final component in the above representation scheme relates to the notion of rationale behind a particular artefact. Rationale is encoded to provide a reasoning for selecting amongst alternatives. Such knowledge is captured by the predicate *rationale* which is defined as follows:

rationale(ListofTerms)

ListofTerms is (symbol,PreferenceExpression) and PreferenceExpression is defined either as preference clause (e.g., p(x,y)) or indifference clause (i(p,z)) where p(x,y) implies that x is preferred to y, while i(p,z) implies that p and z are indifferent.

The notion of preference expressions is used in the present work to provide a logical way for reasoning about design alternatives (Akoumianakis and Stephanidis, 1997a). For example, in the specification of Table 6.1, the deposited preference and indifference expressions for the attribute inputDevice suggest a

---

[6] It should be noted, however, that these details are available anyway as resource files.

**Table 6.1** Internal specification of recommendations for the resizing toolbar; an example of knowledge representation

```
1.  specification ( windowNavigation, resizeToolbar, switchAccess,
2.      [  container,
3.         [  attribute(general, α(accessPolicy, keyboardEmulator, []),
4.            attribute(general, α(topologyPolicy, horizontal, []),
5.            attribute(general, α(inputDevice, twoSwitchScan, []),
6.            attribute(general, α(inputTechnique, scanning2D,
7.               [  parameter(selectionSet, symbol,[]),
8.                  parameter(scanmode, 1,[]),
9.                  parameter(timeScan, 100,[]),
10.                 parameter(dm_bordercolor.red, 1, []),
11.                 parameter(dm_bordercolor.green, 1, []),
12.                 parameter(dm_bordercolor.blue, 1, []),
13.                 parameter(dm_borderstyle, 2, []),
14.                 parameter(dm_borderwidth, 1, []),
15.                 parameter(em_borderwidth, 3, []),
16.                 parameter(em_bordercolor.red, 1, []),
17.                 parameter(em_bordercolor.green, 10, []),
18.                 parameter(em_bordercolor.blue, 255,[]),
19.                 parameter(em_borderstyle, 1, [])
20.              ]),
21.           attribute(general, α(outputDevice, VDU, [])
22.        ]
23.        button,
24.        [  attribute(presentation, α(label.font, helvetica, []),
25.           attribute(presentation, α(label.size, 12, []),
26.           ...
27.        ],...
28. rationale(
29.        [  (inputDevice,                  [
           p(1_switchScan,2_switchScan),
30.
               i(2_switchScan,5_switchScan)
31.                                          ]),
32.           (navigationPolicy,             [
              p(automatic,explicitActivation)
33.                                          ]),
34.           ...
35.        ])
36. )
```

ranking for the alternatives (i.e., 1switchScan, 2switchScan and 5switchScan) into two indifference classes, namely:

1st indifference class: <1switchScan>

2nd indifference class: <2switchScan,5switchScan>

It is important to mention that the preference and indifference expressions in the definition of the predicate rationale are both bound to an assigned criterion which, in our example, is switchAccess (line 1 in Table 6.1). It should also be mentioned that preference profiles comprising preference and indifference expressions are likely to vary from one design case to another. This variation may be accounted for by the designer when updating a particular design case.

### 6.4.4 Propagation of recommendations

Propagation of the recommendations to user interface implementation is facilitated by interrogating the design repository. In other words, when the programmer implements the user interface, dedicated facilities of the Application Programming Interface (API) may be utilised so as to encode explicitly the interrogation of the repository—this allows the run-time libraries of the user interface development toolkit to acquire a project's recommendations and apply them to the interface being implemented. An interrogation cycle involves passing data to the repository regarding the *context* of an interaction and, subsequently, the delivery of *design recommendations* that can be directly embedded into the implementation of a user interface. In this case, the design repository delivers the recommendations in a toolkit-specific format (see Figure 6.6) through message-passing. It should be noted that such a format can be revised to suit different toolkits. In our case, the format depicted in Figure 6.6 was devised for the purposes of the development toolkits used in the project.

The first line, in the specification depicted in Figure 6.6, is a counter giving a total of the recommendations generated for a particular interaction platform. In this case, it is specified that 430 recommendations were compiled in total. Following this line, the recommendations are listed by task context identifier (i.e., Run_Visual_Keyboard, Run_Training_Module) and interaction object class (i.e., DeskTop_FrameWindow). It can be seen that the compiled recommendations cover several attributes of lexical interaction as supported by the target toolkit. Thus, the selected device is single switch with time scan (from a range including touch tablet, keyboard, mouse and joystick), the output device is the screen, while the remaining recommendations specify parameters of the scanning technique used.

The design repository has been implemented as a server under the WINDOWS 95™ environment. User interface toolkits may interrogate the server using dynamic data exchange (DDE) by passing information on the current project and its parameters (i.e., the interaction metaphor, the task context, the interaction object class and the attribute of the interaction object class being considered), so as to establish the discussion topic. This simple protocol of communication is schematically depicted in Figure 6.7, which shows a request for all

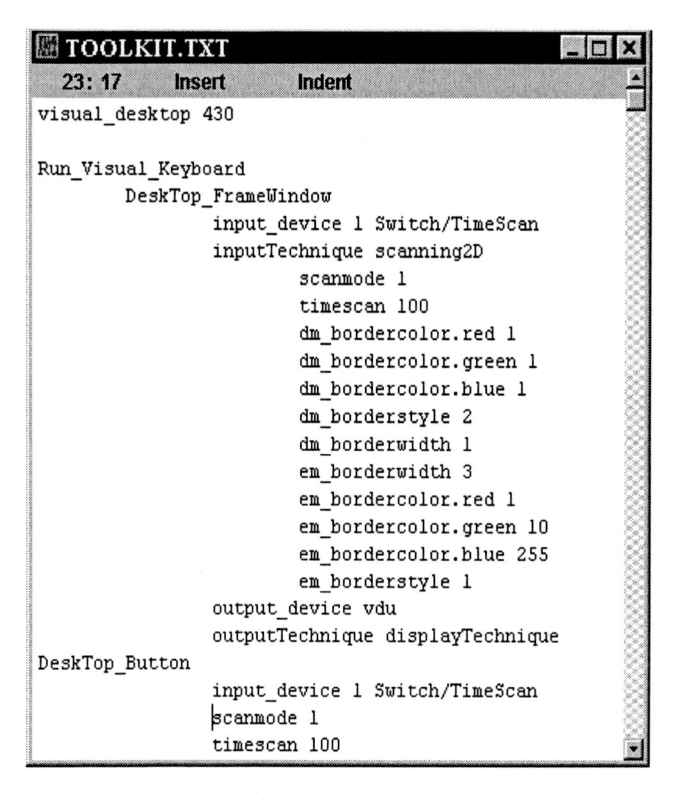
TOOLKIT.TXT
23: 17 Insert Indent

```
visual_desktop 430

Run_Visual_Keyboard
        DeskTop_FrameWindow
                input_device 1 Switch/TimeScan
                inputTechnique scanning2D
                        scanmode 1
                        timescan 100
                        dm_bordercolor.red 1
                        dm_bordercolor.green 1
                        dm_bordercolor.blue 1
                        dm_borderstyle 2
                        dm_borderwidth 1
                        em_borderwidth 3
                        em_bordercolor.red 1
                        em_bordercolor.green 10
                        em_bordercolor.blue 255
                        em_borderstyle 1
                output_device vdu
                outputTechnique displayTechnique
DeskTop_Button
                input_device 1 Switch/TimeScan
                scanmode 1
                timescan 100
```

**Figure 6.6** Extract from design recommendations delivered

recommendations for a particular object class (e.g., button) engaged in the task context (e.g., Run_Visual_Keyboard). It should be noted, however, that the design repository may provide alternative formats for these recommendations so long as the semantics remain the same.

As a concluding remark, it is important to mention that the above protocols for inter-operation between the repository and the run time libraries of the user interface development toolkit assume certain extensions in the API of the toolkit. Specifically, the API should provide a dedicated function which allows the developer to utilise externally generated design recommendations, such as the above, automatically. Although the details in which this is achieved are beyond the scope of this chapter, it should be mentioned that such toolkits have been developed (Savidis, Stergiou *et al.*, 1997; Savidis, Vernardos *et al.*, 1997) and have been used by the ACCESS consortium partners (see Acknowledgements) to develop user interfaces in two selected application domains (see Section 6.5).

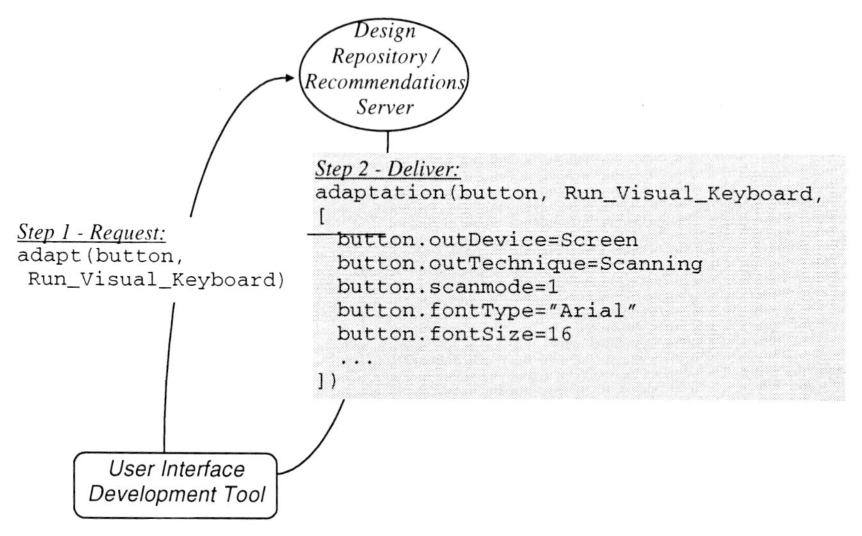


**Figure 6.7** Sharing the recommendations

## 6.5 DISCUSSION: APPLICATION EXPERIENCE AND LESSONS LEARNED

Accessibility recommendations of the type and format discussed in this chapter were developed, in the context of the ACCESS project (see Acknowledgements), for two application domains, namely, the development of an educational hypermedia application accessible by blind users and the development of two interpersonal communication aids for language–cognitive and speech–motor-impaired users. In this context, the objective was the collection of a critical mass of design recommendations which would subsequently grow as the developer's experience grows towards experience-based accessibility recommendations, thus facilitating a living design memory of accessibility.

The two toolkits for which the recommendations have been derived were HKTOOL (Savidis, Stergiou *et al.,* 1997) for the non-visual environment and an enhanced version of the Microsoft Foundation Class library with embedded scanning, called SCANLIB (Savidis, Vernardos *et al.,* 1997) for the communication aid applications. Both toolkits were generated by the PIM tool (Savidis, Stephanidis *et al.,* 1997), which provided a uniform programming layer across the two platforms. In both cases, a dedicated function, namely *adapt* of the API of the toolkits, was utilised by interface developers to interpret and embed (i.e., automatically apply) the accessibility recommendations into user interface implementations. The resulting prototype systems were subject to usability evaluation with end-users yielding very satisfactory results (Kouroupetroglou *et al.,* 1996; Petrie *et al.*, 1997).

In this endeavour, a number of challenges were encountered which revealed the complexity involved when designing for accessibility. Some of these challenges were related to the introduction of new tools for user interface development and the fact that neither designers nor developers were accustomed to building interactive software with such high-level tools. This was further complicated by the late availability of sufficient documentation, in the form of styleguides or programmer's reference manuals, which increased the time and effort required for users of the tools to familiarise themselves with the underpinning concepts.

In addition to the above, several challenges were related to the selection and interpretation of guidelines in each application domain. First of all, the guidelines relevant to the two application domains were radically different with respect to their nature and source. For example, in the case of the interpersonal communication aids, the guidelines which were considered relevant reflected both syntactic (i.e., dialogue for scanning-based interaction) aspects of the interface, as well as physical organisation of on-screen interaction elements (i.e., toolbar location and topology, graphical elements, text size and fonts, etc.). On the other hand, the non-visual guidelines reflected primarily navigation-oriented concerns (i.e., how the user navigates within the overall interactive space, how the user initiates dialogue with a container object, etc.), as well as input and output device considerations. The implication of this variety was not only an increased effort in interpreting the guidelines, for it entailed a representation scheme that would be sufficiently rich to encompass the semantics of recommendations at various levels, including the syntactic as well as the lexical level of the interaction.

A second related problem was the fact that practitioners across the two application domains frequently shared alternative views with regard to the implications of a particular guideline. This necessitated increased communication amongst partners to clarify the context of the guidelines and to define precisely their scope. To address this problem, experts within the consortium were asked to develop experience-based proposals of application and task-context specific recommendations, based on experimental evidence or on the available design wisdom (Smith, 1996; Nordic Cooperation on Disability, 1998; International Organisation for Standardisation, 2000), and relate them to interaction components (i.e., dialogues, object classes and attributes), as provided by the two toolkits. These proposals were subsequently validated and refined so as to depict a common ground for subsequent design activities. In this manner, it was made possible precisely to define the relevant interaction components (such as the virtual keyboard) and their attributes (i.e., the attributes of scanning that would be considered) and to provide the rationale underpinning their existence.

Another problem that was encountered was related to the availability and sufficiency of the empirically validated preference data with regard to the various interaction components. As the available guidelines, in the majority of cases, were grounded on intuition and expert opinion as opposed to experimental evidence, reliable data to derive preference profiles and develop the rationale as discussed in this chapter were totally missing. As a result, it was deemed appropriate to collect such information through end-user consultation. For our three design case studies, this was largely based on a questionnaire survey which was answered by seven disabled end-users in each application scenario.

From the above, it follows that developing a critical mass of design recommendations from the existing pool of accessibility guidelines was time consuming as well as an interaction- and collaboration-intensive effort. Due to these peculiarities, it is considered important to provide mechanisms, such as the one described in this paper, through which experience gained in one design case can be both reused in new design problems and expanded in a way that provides traces of an organisation's accumulated experience and practice. This not only helps the organisation in developing and maintaining a persistent accessibility knowledge base, but also helps individual designers or design teams likely to encounter the same or similar problems in the future.

## 6.6 SUMMARY AND CONCLUSIONS

This chapter has presented the motivating rationale for the development of a method and a supporting tool environment to facilitate the management of accessibility guidelines during user interface development. In particular, we have presented how guidelines may be progressively and incrementally interpreted to provide platform-specific recommendations that can be automatically applied by user interface development toolkits. The main conclusions can be summarised as follows: software tools can provide computational support so as to facilitate and augment the human factors design input delivered by accessibility guideline reference manuals. However, for such input to be effective, several extensions are needed in the prevailing design and development practices. These include: (a) more analytical perspective into HCI design, seeking to envision and unfold design alternatives rather than mere adaptation; and (b) the provision of toolkits that have augmented capabilities, both with regard to interaction facilities offered as well as the APIs supported. Equally important is the capability to reuse past experience to inform future practices. To this end, design support platforms should increasingly offer the means for sharing knowledge as well as accessing, reflecting upon and modifying past cases to facilitate informed and consistent designs.

## ACKNOWLEDGEMENTS

Part of this work was carried out in the context of the TIDE–ACCESS (TP 1001) project Development Platform for Unified Access to Enabling Environments partially funded by the European Commission (DG XIII). The ACCESS Consortium comprised the following organisations: CNR–IROE (Italy)—Prime contractor; ICS–FORTH (Greece); University of Hertfordshire (United Kingdom); University of Athens (Greece); NAWH (Finland); VTT (Finland); Hereward College (United Kingdom); RNIB (United Kingdom); Seleco (Italy); MA Systems & Control (United Kingdom); PIKOMED (Finland).

CHAPTER SEVEN

# Managing HCI Guidelines with Hypertext on the WWW

Luc Goffinet and Monique Noirhomme-Fraiture

## 7.1 INTRODUCTION

Having on-line cross-referenced HCI guidelines would of course be helpful to software designers, as guidelines dealing with the same aspects of HCI for people with disabilities are currently scattered among different sources, making it hard for people to find relevant advice in the field (Stephanidis, 1997). Hypertext cross-references within different documents on the WWW covering the same area would also be a major improvement. These are the issues that we shall try to address in this chapter.

The aims of this chapter are mainly twofold. Firstly it concentrates on techniques to gather human–computer interface guidelines for people with disabilities from different sources, mainly on the WWW. Secondly it deals with the problem of merging these guidelines and converting them into hypertext on the WWW.

The method that we have used to generate hypertext relies on statistical measures to generate cross-references. We shall briefly present a case study concerning the hypertext conversion of general HCI guidelines. Finally we shall tackle the problem of specific HCI disability guidelines, by showing how four different sets of guidelines have been integrated into a single on-line hypertext database.

## 7.2 GATHERING HCI GUIDELINES

Guidelines for human–computer interaction are abundant in the scientific literature. Famous examples include Smith and Mosier (1986) and more recently the US Nuclear Regulatory Commission. HCI guidelines also exist for people with disabilities, though more sparingly. Here are some of them: Vanderheiden and Lee (1988), Microsoft Corporation (1995), McCann (1997), Trace Center (1997) and Nordic Cooperation on Disability (1998). European research projects such as INCLUDE or human factors institutes such as the HUSAT Research Institute at Loughborough University aim to bring together the pieces of the jigsaw puzzle.

Up to now, most of these guidelines are available in electronic form and some of them even exist in a hypertext format (Vanderdonckt, 1995). However, to our knowledge, very little concerning HCI and people with disabilities has yet been turned into full-blown hypertext, except for perhaps the INCLUDE project about public access terminals (Gill, 1997 and Chapter 14 in this volume).

We shall address the issue of retrieving existing guidelines scattered on the WWW.

### 7.2.1 Current search engines

The use of Internet search engines such as AltaVista or Yahoo to track down guidelines is valuable; however, the manual procedure to extract relevant information from the WWW is somewhat limited. No automatic detection of new guidelines can be made and the signal-to-noise rate is poor. Furthermore, the rating of references returned by search engines is not taken into account by the searching process. So, every time that interesting information is retrieved the whole process has to be repeated the next time a similar search is required.

### 7.2.2 Improving information retrieval on the WWW

We have thus started to design a new tool for retrieving information on the Web. It is based on a three-step method. Let us take as an example a search for guidelines concerning the blind and the use of GIF (graphic file format) pictures on the WWW. The keywords for this query should be: *guideline, blind, gif, picture.*

#### *7.2.2.1 Use of meta-search engines*

Our query (four words) is first sent to a meta-search engine. Generally such engines provide an all-in-one searching mechanism. Instead of querying all search engines separately, they are all queried in one go. In an ideal world, the results should be sorted by relevance and merged by the meta-search engine. However, few of them perform well, so our method requires an additional step.

#### *7.2.2.2 Sorting and merging references*

The references returned by the meta-search engines are retrieved and the corresponding documents on the WWW are fetched. Keywords within them are compared to the keywords used in our query so that we can sort the documents according to their relevance to the original query. A document containing all our keywords, several times, should be given the highest score and appear at the top of the list. As a side-effect, double references can also be deleted.

#### *7.2.2.3 Relevance feedback*

To improve the efficiency of our information-retrieval process, the expert (the one who has constructed the query) has to assess the relevance of the information retrieved by the search engine. By taking a quick look at the list returned in the second step, and by casting a yes/no vote with respect to the relevance of the documents, he or she can provide the system with some insight into which documents are relevant and those that are not. With such information, the system can complete the initial query with new keywords (those that appear frequently in relevant documents and not in irrelevant ones) and add keywords to avoid (those that appear frequently in irrelevant documents but not in the relevant ones). Then the process starts again at the first step. This way it yields an improved version of the reference list.

The query and its results, refined by relevance feedback, can now be saved and thus reused later on. The material gathered with this technique can be considered for inclusion in a general database for HCI guidelines for people with disabilities. We have not yet fully developed the tool that implements our technique but one of its major features would be a permanent watch of the WWW, so that when new documents appear that match a query, these ones are automatically inserted and linked to the existing database.

## 7.3 FROM TEXT TO HYPERTEXT

After gathering different interesting guidelines on the Web (mainly from text files retrieved thanks to meta-search engines as above), we are faced with the problem of placing all these pieces of text inside a common and new hypertext.

A major problem is how to structure the data gathered above. A manual structure would ruin all our attempts at automatically adding new guidelines to our corpus. So a free structure that is accessed with keywords seems the best way to proceed. The building of indexes to access this structure would be manual, whereas the linking of documents from the index and between one another would be mainly automatic.

The problem of converting text into hypertext is mainly twofold. The first part of it is the generation of *hierarchical links*, such as 'table of contents' references, 'next', 'previous', 'top of chapter' buttons, etc. This is of course the easiest part of the translation process. For example, generating several HTML (Hypertext Mark-up Language) documents from a single well-structured Word document only requires a good set of macros. More generally the problem of turning regularly structured text into hypertext has already been addressed by many researchers (Furuta *et al.*, 1989; Sarre and Güntzer, 1991; Salminen *et al.*, 1995).

The second part of the conversion process is the automatic generation of cross-reference links without any textual clues (such as 'see also' or 'cf.'). So far this has largely been done by hand, by people who have sufficient knowledge about the domain and who can point out *semantic links* between different parts of a document. Automatic cross-referencing is a challenging task, because hypertexts tend to grow in size and this task is highly time-consuming.

After a brief introduction to the methods that we have used to generate hypertext (thanks to a real-world case study), we shall then tackle the specific problem of turning HCI guidelines into hypertext.

### 7.3.1 Generation of hierarchical links

The most obvious part of the hypertext conversion process is thus the generation of a set of HTML files, all related to one another according to their spatial order (first, next, previous, last, top, bottom). All that is required as input to this process is a well-structured source text, where the hierarchical structure is marked by styles attached to paragraphs (such as ‘Heading 1’, ‘Heading 2’, etc., in a word processor).

We have used the RTFtoHTML software (Hector, 2000) to translate the text from the well-known RTF format into a hierarchy of well-structured HTML files. This also implies that the output of the process always has the same structure, which ensures consistency across different guideline sets.

### 7.3.2 Generation of semantic links

The remaining task is to generate meaningful cross-reference links that can help people gather useful knowledge about related items without browsing through the entire collection of guidelines.

There are in fact several approaches to detect cross-reference links. The first one is the *syntactic approach*. Cross-references can be detected by occurrences of phrases such as: *“see also...”, “this can be likened to ...”, “in ... Mister X says that”,* or by more sophisticated mechanisms to detect quotations (Conrad, 1994). This approach is a first step towards the final goal. In the case of Smith and Mosier’s guide, described in the case study below, it has been followed by Vanderdonckt (1995) to automatically generate a hypertext for Windows. This approach can be extended by *natural language analysis*, used by some researchers to detect hypertext ‘anchors’ (Lehnert, 1992).

The second approach comes from the artificial intelligence field. It is based on *neural networks*. As the description of this approach is beyond the scope of this chapter, the interested reader is referred to Kwok (1995) and Ueno and Ogawa (1993).

The third approach is based on *statistical methods* used in information retrieval. When one searches for a document in a database, a keyword vector is generated from the query and compared to the keyword vectors representing the documents inside the database. In a nutshell, a keyword vector consists of any significant word that belongs to a document. A similarity coefficient is then computed for every pair query/document. The best matches, i.e., those documents that have the highest similarity coefficient with the query, are given as output of the process. More details are provided by Salton (1989).

This is the basic idea behind the generation of cross-reference links based on the statistical analysis of a document. Every document in the database is compared with all the others and, when similarity coefficients are above a given threshold, a link can be generated.

## 7.4 A CASE STUDY

We have chosen the guidelines from Smith and Mosier (1986) to carry out our first experiment. A complete report of this study can be found in Goffinet and Noirhomme (1996).

Their guide contains 957 rules for designing more usable software interfaces. Our goal has thus been to find a way to create meaningful *semantic links* between those rules. This work has been made easier by the fact that Smith and Mosier had already inserted 701 such links into their guide. This has been important, not only for evaluation purposes, but also to serve as a basis for optimised keyword weights used in probabilistic information retrieval.

We have chosen the statistical approach described above, using the *weighted keyword vector model* that is commonplace in information retrieval. The main variations in the model are mainly as to which words serve as keywords, how to weight those keywords and how to compute the similarity functions between two vectors.

### 7.4.1 Results

Our tests have shown that the best keyword extraction technique is to use full-text indexing. The best weighting function is the well-known 'Term Frequency * Inverse Document Frequency' which gives a bigger weight to keywords that are rare in the whole set of texts and frequent in a single text. As for the similarity function, it is the cosine function, which gives 1 for similar vectors and 0 for totally different ones.

After the computation of similarity coefficients and retaining the best 1000 links, we have observed a precision rate of 80%, which means that 80% of the generated links have been judged 'relevant'. The recall rate has been 25%, which means that 25% of existing manual cross-references have been retrieved. This result might imply that our techniques are not very efficient, but the high precision rate means that even if we did not retrieve many manual cross-references, those which were generated would still be meaningful.

More sophisticated keyword extraction techniques could be considered. For instance, identifying noun pairs or concepts. However this requires either important domain-dependent knowledge-bases or natural language analysis, which is beyond the scope of our work.

### 7.4.2 Using relevance feedback

In probabilistic information retrieval (PIR), when some knowledge is available about the relevance of a document associated with a query, the keywords can be re-weighted according to this information. This maximises the similarity function for the relevant pairs query/document. The reader interested in PIR will find more information with van Rijsbergen (1979) and Fuhr and Buckley (1991).

Thus, the 701 existing cross-reference links have been used to weight the 4096 keywords that had been identified in the text. We have discovered that there was a major improvement in the recall rate when we used this sort of *relevance feedback mechanism*. More existing cross-references were retrieved (30%), and so the quality of other generated links proved superior to other methods. This has not been fully confirmed, as a precision of 82% had been observed when browsing through the generated links (about 400 out of 700). Yet this result was interesting, as it maximised the retrieval of links whose type had been judged relevant by somebody.

### 7.4.3 A semi-automatic procedure

After the experiment carried out above, we outlined a semi-automatic procedure for cross-referencing in a hypertext. Its first step could be to run an automatic linking algorithm on the set of documents to link (with simple full-text indexing and TF * IDF similarity computation).

The second step would require some human relevance feedback about whether the links generated during the first step are relevant or not. In our case, for example, we have been able to evaluate 400 links out of 700 generated—the more, the better of course.

This could lead to a second running of the linking algorithm where better links could be discovered and implemented automatically.

This opens the path to semi-automated cross-referencing link generation. One will notice that this procedure is very similar to the one used in information retrieval for relevance feedback. As its efficiency has been acknowledged many times in that field, this seems to be the case here too.

## 7.5 DISABILITY GUIDELINES

After gaining initial experience with the process of hypertext conversion and cross-reference generation on a general HCI guidelines text, we have processed other sets of HCI guidelines. Smaller sets of guidelines are indeed available on the WWW for those wishing to make their software more accessible to people with disabilities. We have thus generated on-line hypertext guidelines from the following sources: Vanderheiden and Lee (1988), Microsoft Corporation (1995), Trace Center (1997) and Gill (1997). The result of our hypertext conversion process can be consulted on the WWW at the following address: www.info.fundp.ac.be/httpdocs/guidelines/.

The first step of the process was the generation of a hierarchy of HTML files, all related to one another by hierarchical links. Then we had to address two separate issues. The first one was the cross-referencing within a single set of guidelines (generation of internal links). The second one was the generation of cross-references between different sets of guidelines (generation of external links).

### 7.5.1 The hierarchical links

The starting point for all the conversions into hypertext was ASCII files containing all the guidelines (one file per source). We then added some structure to them using a word processor (separating headings of level 1 from level 2 and so on). The hierarchy we introduced by simply attaching styles to paragraphs was the basis for the generation of a well-structured set of HTML files, each containing a single guideline and related to one another by simple navigational links (next, previous, top, bottom). This task was achieved by a simple RTF to HTML converter.

The main advantage now is that all these sets of guidelines have the same presentation on the WWW. We have introduced consistency in the presentation of all these guidelines from different sources. From now on, we can easily add new sets of guidelines to our database from simple ASCII files. One can appreciate the result in Figure 7.1. It is derived from the guide (Microsoft Corporation, 1995) that we have converted into hypertext.

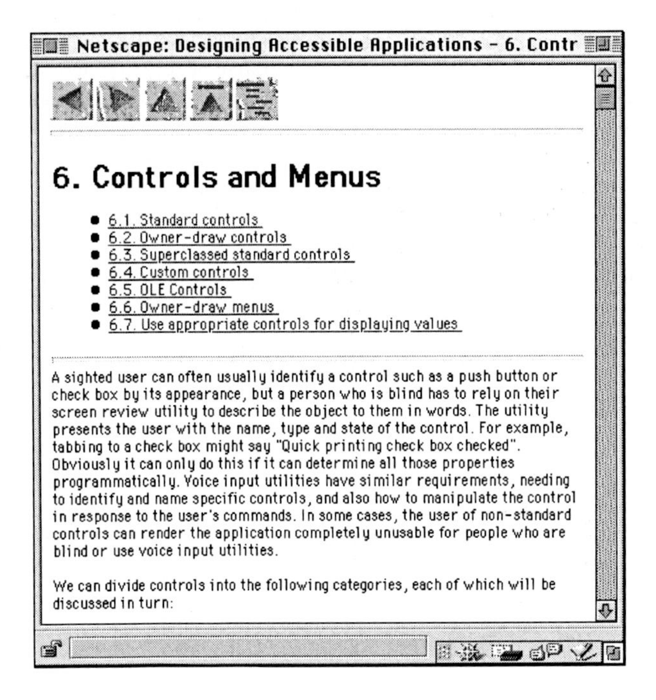


**Figure 7.1** HTML file with hierarchical links

### 7.5.2 The internal links

After the generation of HTML files, we used our algorithms to produce the internal cross-references within given sets of guidelines. A matrix of similarity coefficients between all the guidelines was computed and we retained the best 1% of them to automatically generate cross-reference links. This has produced good results, although, given the relatively small size of each guidelines set, the precision rate is not really significant here.

### 7.5.3 The external links

We also computed similarity coefficients between guidelines belonging to different sets. The idea was to retain the highest similarity coefficients to generate external cross-references.

### 7.5.4 Evaluation

The results of the automatic cross-reference links generation are summed up in Figure 7.2. The four sets of guidelines are in rectangles. The internal links are represented by semicircles, while the external links are straight lines. Lines and circles are labelled with numbers giving the number of links (relevant/not relevant) generated by our technique. The thickness of a line/circle grows with the number of links. There is a total of 84 links (either internal or external), 66 of which have been judged relevant (by ourselves) and 18 not. This is an interesting result as it shows that we have once again reached a precision rate of 80% (this is in fact the *relevance rate* for those not acquainted with the technical terms).

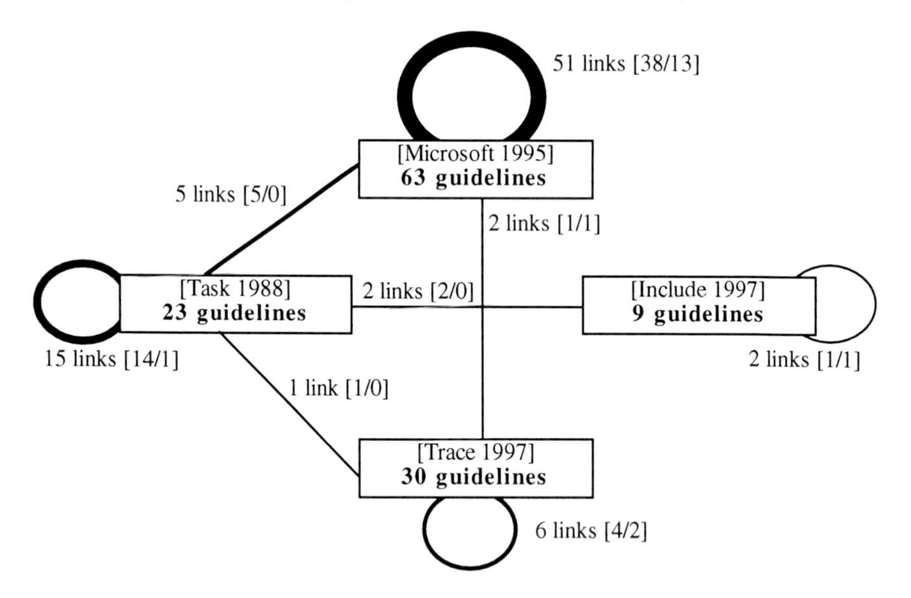


**Figure 7.2** The generated links, both internal and external to our 4 sets of guidelines

There is much that we can learn from this graphic. Firstly, it shows that most of the generated links are internal ones. This might be because people tend to define and use their own vocabulary when they write a guide, thus sharing fewer common words with other people. Or it might be that the guides do not merely address the same issues. This is particularly true with the *Public Access Terminal* guide (Gill, 1997), which deals with hardware devices such as cards, which are not covered by other guides. This indicates that external cross-referencing is much harder than internal cross-referencing.

However, there is a good external link in Figure 7.3. It shows a good example of an automatically generated external link. The similarity coefficient between the two guidelines is 0.16 and one can notice that these guidelines are indeed very close to each other.

The other major problem is the 18 links generated by our algorithm that we rejected as they were not relevant. Here, the main fact is that statistical

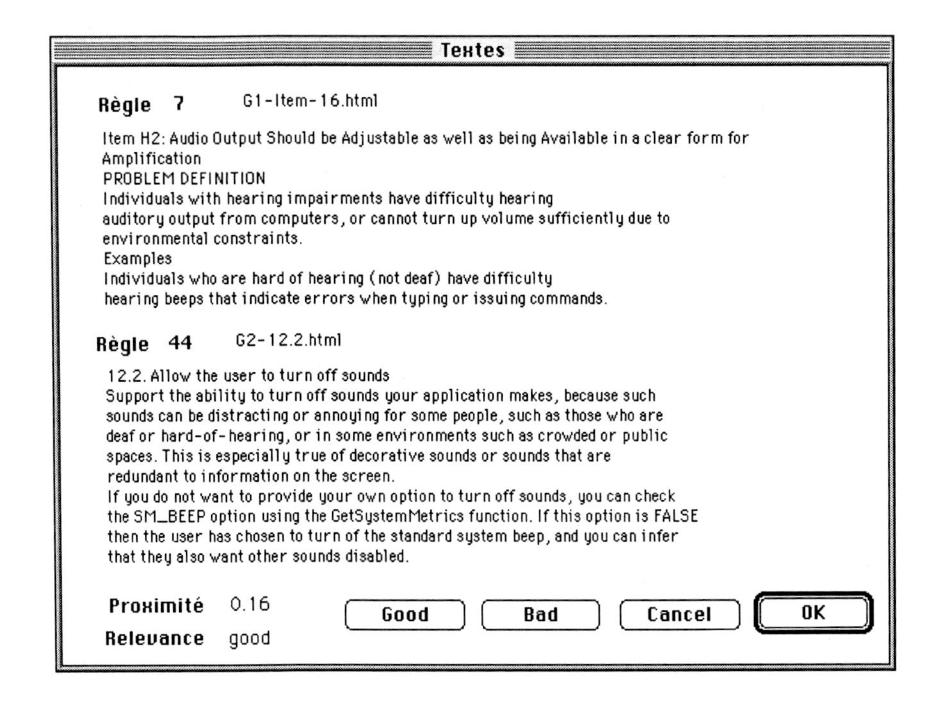


**Figure 7.3** A good external link between two guidelines belonging to different sets

measures alone do not access the meaning of sentences, and thus produce cross-references that are not always relevant. The meaning of a word is of course not 'context free'. In Figure 7.4, there is a sample of a bad external link produced by our algorithm. Although the similarity coefficient is 0.15, one can notice that these guidelines do not address the same topics. The confusion here arises from the use of the same words in different contexts. The word 'control' has (at least) two different meanings, thus introducing confusion into the process of similarity detection. This reveals why a purely automatic algorithm will always yield some bad cross-references that should be avoided in the final hypertext version of HCI guidelines.

## 7.6 CONCLUSIONS

Statistical methods alone cannot generate well-linked hypertexts but they can provide valuable help in the framework of a semi-automated linking process. After an automated first step (the computation of similarity coefficients based on keyword frequencies), relevance feedback can be given to the process to yield better cross-reference links.

A human expert is thus needed. His or her role would also be to organise the guidelines before the conversion process so that they are neither too long nor too

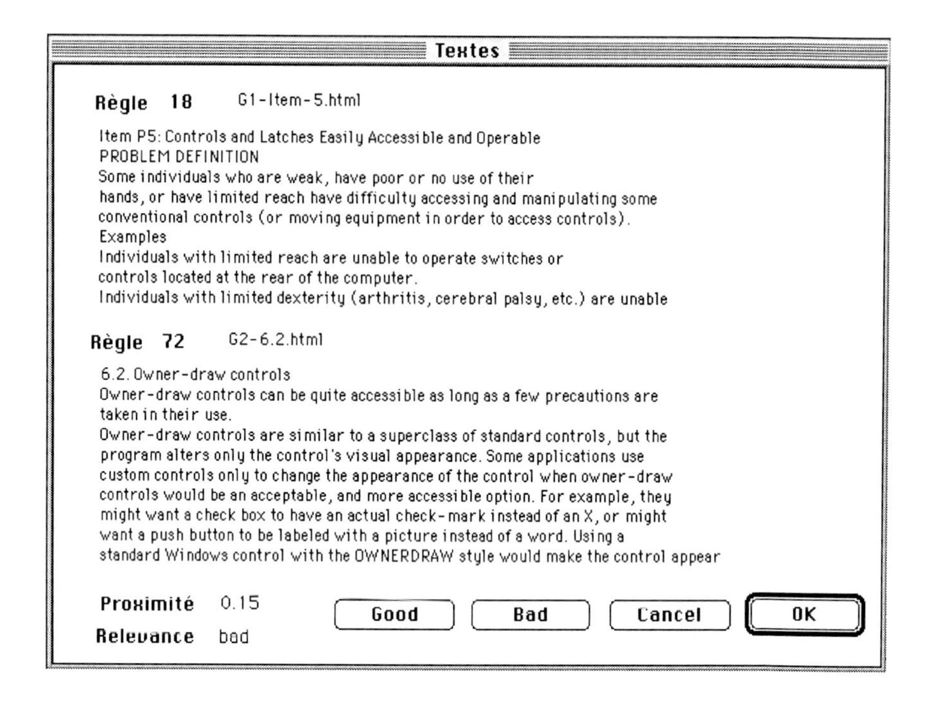


**Figure 7.4** A bad external link between two guidelines belonging to different sets

short, and can be integrated in an existing on-line cross-referenced hypertext database.

We feel that a hypertext database on the WWW, with good cross-reference links, is a valuable tool for Web ergonomics experts who want to advise designers. Our work might help them, should we gather further guidelines to enhance our database (those with guidelines available are requested to contact the authors). In addition it is accessible from all over the world thanks to the Internet.

If we wish to provide such a tool to designers, we feel that they need a more appropriate product: they usually want to find the information quickly and in an understandable form. The guidelines have then to be rewritten in a more uniform way, using the vocabulary of computer scientists, with a uniform structure such as 'title, abstract, extended description, examples, etc.', with keywords and quick access means. The SAPHIR project, supported by the Walloon Region of Belgium, is developing such a product that will be designed in French (de Baenst-Vandenbroucke *et al.*, 1999).

CHAPTER EIGHT

# Bobby: A Validation Tool for Disability Access on the World Wide Web

Charles G. Hitchcock, Jr.

## 8.1 INTRODUCTION

As the World Wide Web becomes central to success in education, business and all aspects of life, it must be made universally accessible. Bobby, an innovative Web-based public service, helps designers worldwide create sites that are usable by everyone, including those with disabilities.

With Bobby, Web designers can learn about barriers and quickly implement changes that provide increased access and ultimately benefit all users. As Web designers create more accessible sites, they learn to think about users' varied needs. This awareness results in sites that are comprehensible to a greater population, including people with age-related barriers (such as failing vision) and those with language, cultural or economic barriers. The impact of this increased accessibility is vast and profound. If the World Wide Web is one day available to everyone, it can provide opportunities for learning, literacy and business that simply would not otherwise exist.

Many of the issues addressed by the *Web Content Accessibility Guidelines* as recommended by the World Wide Web Consortium, or W3C (see www.w3.org/TR/WAI-WEBCONTENT and Chapter 10) apply to anyone using a browser to navigate the Web. Good design for those with disabilities generally leads to unanticipated benefits for those who are temporarily abled. An example of this in the physical world is the benefit of curbcuts for wheelchairs that are widely used by bicycles and parents pushing baby carriages.

A physical connection does not guarantee access to the opportunities the Web offers. For the over 750 million adults and students worldwide with disabilities, including sensory impairments, physical challenges and learning disabilities, the World Wide Web is a challenging place.

For example, highly graphic sites present obstacles for those with visual impairments. For these people, screen readers can translate text to speech but cannot translate graphics. Audio files without text transcripts and videos without captions are inaccessible to those with hearing impairments. Sites that are rich with text provide little information to those with reading difficulties.

Individuals of all ages with disabilities are in danger of being left on the side of the information superhighway. Without careful attention to overcoming the Web's inherent barriers, this significant educational resource will become yet another hindrance to learning for students with disabilities. Like most profound innovations, the Web holds great promise, but it is critical to address the issue of its accessibility or its promise will not be fully realized.

The same applies to workers of all ages who require access to critical job functions now available only through government and corporate intranets and the Internet. Pages that are designed with access in mind make it possible for employees with disabilities to participate and be productive.

## 8.2 BOBBY BACKGROUND

Bobby was developed at the Center for Applied Special Technology (CAST at www.CAST.org) in 1996 in order to provide Web developers with a tool that would educate users about disability access issues on the Web and provide a validation tool to help content authors prepare accessible sites. Bobby grew out of CAST's underlying mission to expand opportunities for people with disabilities through innovative uses of computer technology. To achieve this goal, CAST uses digital media to build universally designed products, that is, educational software, Internet tools and learning models that are usable by everyone, including people of all ages with sensory impairments, physical challenges and learning disabilities.

The first release of Bobby was an on-line tool that parsed the user's HTML (Hypertext Mark-up Language) through the CAST server code and provided a report of Web barriers to the extent that an electronic tool was able to do so. The early versions used *Web Access Guidelines* published by the Trace Research and Development Center at the University of Wisconsin, USA (Vanderheiden *et al.*, 1997 and formerly found at www.trace.wisc.edu but replaced by W3C–WAI guidelines).

Even though a number of organizations offered comprehensive printed guidelines, CAST suspected that Web developers might not actually read them. CAST then came up with the idea of a helpful detective—a Web-based, on-line tool that would expose barriers, encourage compliance with the existing guidelines and teach Web masters about accessibility. Bobby was born.

CAST researchers and Web consultants developed the first version of Bobby in just three months, releasing it in September 1996. The project was funded by general research funds, outside foundation support and royalties from CAST's commercial products. Bobby has since been upgraded every six months to keep pace with rapidly changing Web technology. New versions have included improved page authoring guidelines, new features and technical enhancements. The downloadable application was released for developers of large Web sites who need to test a large number of pages all at once, to test pages before posting them directly to the Web or to test pages behind their security firewalls.

The release of Bobby v.3.0 (August 1998) provided a major upgrade that replaced all previous versions including Bobby 2.0 and 2.01. Both the site and the application offered an improved implementation of the working draft of the W3C's Web Accessibility Initiative (WAI) *Page Authoring Guidelines* and reflected the Page Authoring Working Group's latest revisions (see Chapter 10). The Bobby application could now run under Windows 95/98/NT, Solaris or any Java 1.1 compatible platform.

Major changes to the application included an accessibility summary report for an entire Web site, ability to export all of a Web site's accessibility reports to a set of HTML files, ability to import a list of URLs (uniform resource locators) for accessibility analysis, ability to print accessibility and summary reports, better control over how links are followed in the link finder, and its having been rewritten using the JFC (Java Foundations Classes) to make the application itself accessible to Java-based assistive technologies.

Release of Bobby v3.1 (May 1999) synchronized with the release of the new W3C–WAI *Web Content Accessibility Guidelines*. This release provided a new benchmark for Bobby Approval. Now, in order for a site to become Bobby Approved, it must pass all Priority 1 items in the *Guidelines,* including those that Bobby can check for automatically and those that must be checked by a Web designer manually. A new category in the Bobby report lists those things that must be checked manually and corrected if necessary. At the time of this writing, Bobby is in release 3.2.

### 8.2.1 W3C Guidelines

The Bobby analysis of accessibility is now based on the W3C–WAI *Web Content Accessibility Guidelines*. For example, to become Bobby approved, a Web site must:

- provide text equivalents for all non-text elements (i.e., images, animations, audio, video)
- provide summaries of graphs and charts
- ensure that all information conveyed with color is also available without color
- clearly identify changes in the natural language of a document's text and any text equivalents (e.g., captions) of non-text content
- organize content logically and clearly
- provide alternative content for features (e.g., applets or plug-ins) that may not be supported.

These are just a few examples of the suggestions provided by the W3C. Once a Web site receives a Bobby Approved rating, it is entitled to use a Bobby Approved icon (Figure 8.1).

**Figure 8.1** Bobby Approved icon

### 8.2.2 Other features

An analysis of Web pages for compatibility with various browsers is also provided by Bobby. The analysis is based on documentation from browser vendors when available. Bobby automatically checks sites for compatibility with HTML 4.0. For accessibility and tag compatibility with browser specifications other than HTML 4.0, 'Advanced Options' are provided for testing compatibility with other versions of HTML.

This evaluation is only one step in the process of making a site accessible to as many people as possible. CAST recommends that Web developers use Bobby as a first step to ensure accessible Web page design. There are some important aspects of accessible Web page design that cannot yet be tested by Bobby, so items which must be checked manually are flagged.

While some Web designers seek out Bobby because they are aware of accessibility issues, others discover Bobby when looking for a tool to test how various browsers will handle their page authoring code and how long their sites will take to load. They then discover the significant issues surrounding accessibility. This alternative entry point has enabled Bobby to educate Web designers who were not yet aware of accessibility issues.

## 8.3 BOBBY'S TECHNICAL STRUCTURE AND DEVELOPMENT

Technical notes prepared by Josh Krieger, Bobby's original developer/ programmer were used as a primary source for this section, although it is not intended to provide a comprehensive review of the technical development and structure of Bobby.

Bobby is able to find relatively simple accessibility problems in HTML pages such as missing ALT text for images. It also has a list of all the acceptable HTML elements for a variety of browsers and can check the elements on a page for adherence to this. In addition it will determine the approximate download time for all the images, applets and objects on a Web page as if they were viewed using a slow modem.

CAST originally set out to create a simple Perl/CGI (common gateway interface written in Perl) script that would sit on a Web server and analyze pages for accessibility (or rather conformance with the Trace guidelines). This initial version was developed very quickly and, just near the end, the developer realized that we could not only have an accessibility report, but also redisplay and annotate the analyzed Web page, marking precisely where accessibility errors had occurred.

CAST received considerable feedback about Bobby and we continued to refine the accessibility-checking mechanism and interface. We released the Perl code so that developers could set up their own intranet Bobby, mirror sites, or just use the script from the command-line. Initially, Bobby ran on the CAST ISP (Internet service provider), but it became so slow at times as to become almost unusable. Simultaneously there was much demand for an application version of Bobby. Some organizations and government agencies were understandably reluctant to issue internal pages to the Bobby server located at CAST. The original Perl script could only run under UNIX, and we needed a tool that could run cross-platform, analyze whole Web sites for accessibility and act as both a Bobby server and a Bobby application.

In the summer of 1997, CAST purchased a 200 MHz NT server to run Bobby and began work on translating Bobby into Java. Initially the Java-based Bobby that ran server-side was problematic and it was necessary to turn Bobby into a thin Web server itself rather than running it through some sort of CGI process.

A Java GUI (graphical user interface) front-end was added using the Java 1.1 and the AWT (Abstract Windowing Toolkit) GUI interface toolkit. When this version was finally released in the winter of 1998, it implemented the latest accessibility guidelines that had moved from the Trace R & D Center's care to the W3C's Web Accessibility Initiative.

Time was spent rewriting the AWT Bobby using the newly released JFC (Java Foundation Classes). This finally culminated in mid-summer 1998, just before the original developer/programmer left CAST to further his studies, with a release of a JFC Bobby that was much more stable and a better implementation of the latest WAI *Page Authoring Guidelines*.

### 8.3.1 Technical overview

The Bobby class is the workhorse of the core Bobby package. It is responsible for taking an input stream of HTML, parsing it, and analyzing it for accessibility, HTML compatibility and download time statistics. For the most part, the Bobby class stores the data but does not output it—essentially creating a model/view separation. Accessibility reports can be output using the BobbyReports class.

### 8.3.2 The processing flow

The following simple example illustrates how Bobby analyses a URL:

```
public class bobbyTest {
browserSet bs;
public void bobbyTest()
  {
    HSSElementSet hes = new HSSElementSet();
    try {
      InputStream is = utility.getFileResourceAsStream(browserFileName);
      if (!hes.asciiLoad(new InputStreamReader(is))) {
        System.err.println("Unable to load browser: " + browserName);
        return;
      }
      bs.addBrowser(hes);
    } catch (IOException ioe ) { }
  }
  public Bobby runBobby(URL u) throws IOException
  {
    URLConnection uc = u.openConnection();
    Bobby bobby = new Bobby(bs);
```

```
        Reader ir = new InputStreamReader(uc.getInputStream());
        bobby.run(u,ir,true,true);
        ir.close();
        return bobby;
    }
}
```

### 8.3.3 Bobby run and accessibility analysis

Call Bobby.run() is used to start accessibility analysis. The Bobby.run() module will start an HTMLParser and call two other methods: analyzeTag() and analyzeText(). The analyzeTag() method is responsible for finding accessibility errors in HTML tags while the analyzeText() method finds accessibility errors that occur within the text (i.e., using ASCII art is problematic for people who are blind). When the parser has gone through the whole document, the analyzeWholeDocument() method is called.

When an accessibility error is found, the Bobby.addError() method is called and the error is added to the category in which it belongs: error, recommendation, question, tip or a browser error. Each of these is a class of type errorList. Using the getAccessErrs(), getRecommendations(), getTips(), getQuestions() and getBrowserErrs() methods, and the appropriate methods in the errorList class, all of the error information can be acquired manually by a program. In this way presentation of the accessibility errors are totally separated from their representation. Unfortunately, this is not 100% true because the on-line version of Bobby requires that we insert a little 'bobby' helmet (as worn by British policemen) on the redisplay of a Web page wherever an accessibility error occurs. The code to handle this exception is complex.

### 8.3.4 The Bobby accessibility guidelines

The WAI accessibility guidelines are represented in an XML (Extensible Mark-up Language) file for use by Bobby. This XML file contains information about the priority of errors and how well Bobby can support them (so that Bobby can decide whether to make them errors, recommendations, questions or tips), as well as text descriptions of how to fix the errors for Web site developers using the program. Before a Bobby release is created, these guidelines are compiled and saved in a binary file and the text descriptions are turned into a set of HTML files that are included in the Bobby distribution. The guideline file is loaded statically from Bobby class using the Bobby.getGuidelines() method. The guidelines are stored in the guidelines file and all of their respective information can be accessed through this class.

It is important to understand the theory behind the implementation of Bobby's accessibility guidelines. Each guideline is tied to a particular browser for which it is relevant. That is, a browser is the technology that a blind person would use to access the Web and it, along with the authored HTML, and an assistive technology (i.e., screen reader) determines whether a page is ultimately accessible. The WAI guidelines try to divorce authoring from the browsing technology—and they do a pretty good job—but ultimately they ignore a number of small details that are very important. For example, consider the following entry for image ALT text:

```
<guideline id="g9" created="05/28/98" audience="general" clarity="objective">
<title>Provide alternative text for all images</title>
<wai version="072298" rating="p1" corresponds="a.1.1;a.4.1"/>
<relevance browser="HTML 2.0"/>
<relevance browser="HTML 3.2"/>
<relevance browser="HTML 4.0"/>
<relevance browser="Navigator 1.1"/>
<relevance browser="Navigator 2.0"/>
<relevance browser="Navigator 3.0"/>
<relevance browser="Navigator 4.0"/>
<relevance browser="Navigator 4.5"/>
<relevance browser="Explorer 2.0"/>
<relevance browser="Explorer 3.0"/>
<relevance browser="Explorer 4.0"/>
<relevance browser="Explorer 5.0"/>
<relevance browser="Lynx 2.5"/>
<relevance browser="Lynx 2.6"/>
<relevance browser="Lynx 2.7"/>
<relevance browser="AOL 2.5"/>
<relevance browser="AOL 2.6"/>
<relevance browser="AOL 2.7"/>
<relevance browser="AOL 3.0"/>
<relevance browser="AOL 4.0"/>
<relevance browser="Mosaic 2.1.1"/>
<bobby support="full" easetofix="easy"/>
<description><![CDATA..............]
```

Notice that ALT text is only required for browsers that support ALT text. For example, WebTV did not support ALT text and hence it was not included. The WAI guidelines would say, "include ALT text always". But Bobby says, "know your audience and only include ALT text when necessary". In order to make a Web site accessible to a large audience, one analyzes it for accessibility against a large number of browsers. Subsequently, the Bobby Approved rating is defined as being accessible under four browsers: Navigator 4.0, Explorer 4.01, Lynx 2.7 and HTML 4.0.

### 8.3.5 The Bobby GUI application

The Bobby GUI application is the most complicated piece of the Bobby system. Built using the Java Foundation Classes (JFC/Swing), it allows for a full cross-platform GUI that, because of the JFC, will morph itself into the look and feel of the user interface of the platform on which it runs.

The GUI itself has been made as accessible as possible to assistive technologies in the following ways. A command-line interface is provided for the assistive technologies available today. The application itself is written in JFC which means that all of the GUI components are by default accessible.

- Image descriptions were added for images and image buttons
- Keyboard accelerators were added to menu items
- Form controls were grouped with their labels
- Code was internationalized
- Mnemonics were added to the menus so that they can be pulled up from the keyboard (i.e., ALT–F to pull-down the File menu)
- A command line switch -j was added to Bobby so that the GUI could be started up using the JFC JEditorPane component, rather than the HotJavaBean to display Bobby reports and the Frequently Asked Questions (FAQ). This is very important because the JEditorPane component is fully accessible to assistive technologies, whereas the HotJavaBean is not.

### 8.3.6 The command line interface

All of the functionality of the GUI interface is available through a command line interface. The purpose of this interface is twofold: it provides a convenient method for batch processing and it provides an accessible solution for people using today's screen readers.

There are two programs in the command line interface: linkfinder and bobbycl. The linkfinder program traverses links on a Web site and creates a file that contains a list of URLs. The list of URLs is then given to the bobbycl program that outputs an entire directory of HTML files containing the accessibility analysis of these HTML files.

### 8.3.7 Compilers, libraries and additional tools

Version 3.0 of Bobby was developed with Symantec Visual Cafe Pro Database Edition 2.0 as the main IDE (integrated development environment).

Additional tools used include:

- Perl for Win32
- Sun Java Development Kit 1.1.6
- Microsoft Java SDK and VM 3.0
- IBM's Java-based XML parser

Other tools which were helpful include:

- Emacs for NT
- Bash for NT.

## 8.4 HOW BOBBY IS USED

Bobby (www.cast.org/bobby) is an efficient, easy-to-use, free public service that helps Web developers make their pages accessible. Users simply type in the URL of the Web page they want to analyze and, in seconds, Bobby delivers a full accessibility report. Bobby gives a detailed analysis, specifying each barrier and explaining how to eliminate it. A site that is deemed accessible can then display the Bobby Approved icon.

For example, if Bobby finds images that do not have essential supplemental text descriptions (alt tags), it highlights the errors and specifies the remedy. Like its British law enforcement namesake, Bobby is meant to help, not reprimand; it provides prioritized suggestions for making sites more accessible.

Bobby has revolutionised the way many Web designers respond to accessibility. Web developers are unlikely to study, follow and implement complex printed guidelines. CAST therefore developed Bobby—an on-line tool—to teach Web developers how to put these standards into practice by providing specific feedback on their designs, either during development or afterwards.

### 8.4.1 Bobby as a teaching tool

Bobby teaches Web designers to correct access problems in a simple and inviting way. Users type in the URL they want to analyze and Bobby tests the page; it highlights problems, gives an accessibility rating and explains any access problems found. Bobby also gives links to other sites that discuss access issues.

Because Bobby is a highly effective teaching tool, Web designers (including the rapidly growing number of non-professional designers who are putting up sites) learn skills they can apply later. The more people use Bobby, the more likely they will be to integrate accessibility principles into their way of thinking and their style of Web site design.

### 8.4.2 Bobby as an advocacy tool

Bobby has encouraged further research and creativity worldwide by Web designers who seek to ensure the widest possible audience to their sites. Bobby is contributing to increased awareness, the cornerstone to making the Web a useful tool for all users. By promoting accessibility, Bobby is changing how Web developers design their sites using what CAST terms a more universally designed approach. As more developers use Bobby, accessibility on the Web will no longer be an afterthought—rather, it will become a fundamental part of the design process.

### 8.4.3 Defining problems, developing technologies

Bobby is unique because it uses existing technology to address a significant problem inherent within the emerging Web. The use of new multimedia Web technologies has made the Web an incredibly dynamic environment—but also an increasingly inaccessible one to people with disabilities. CAST has recognized the need for a proactive strategy for implementing the guidelines. Bobby provides an online tool that enhances other Web accessibility efforts by showing Web designers how to implement the W3C printed guidelines—it extends their reach.

This hands-on, pedagogical approach makes Bobby unique. Because the program instantly analyzes Web pages online, it gives users immediate feedback; because it is Web-based, it is available to anyone in the world who has access to a browser and an Internet connection.

Bobby's creative, contextualized reporting system displays error icons precisely where barriers occur. This unique approach makes the program an important learning tool for designers and has provided accessibility advocates with new ways of explaining access problems.

Bobby also employs the latest innovations in Web technology. Written in the Java programming language, the downloadable version can run on many different hardware platforms. More importantly, Bobby uses Java's most current accessibility features, making the program itself accessible to developers and advocates who have disabilities themselves.

Web developers often ask if authoring accessible Web sites will require mostly text and only limited use of media and advanced technologies for creating interactive and dynamic sites. Further, they wonder if the development of new Web-based services for business, health, education and entertainment will become uniform in nature due to the constraints required to comply with Web accessibility guidelines. Although some existing technologies have presented barriers, in most cases other technologies which are, or can be made, accessible may be used to offer animated, interactive and media-rich Web sites. Of particular note, SMIL (Synchronized Multimedia Integration Language) is a language for describing interactive synchronized multimedia distributed on the Web, see www.w3.org/TR/REC-smil/ It is XML-compliant with a syntax similar to HTML. SMIL can provide timing, synchronization, layout, media presentation and hyperlink support and can be made fully accessible. W3C's commitment for advanced technologies bodes well for the future of a complex yet accessible Web.

## 8.5 BOBBY'S FUTURE

CAST maintains and updates Bobby regularly to keep it functional and current. We also plan to provide enhanced, ongoing technical support as Web developers make their sites more accessible. Through its corporate sponsorships and usual fund-raising channels, CAST intends to continue offering Bobby as a free, downloadable public service.

Our aim is to ensure that Bobby meets the rapidly changing needs of Web designers. Currently, CAST is working with IBM, Microsoft and Sun Microsystems to guarantee that Bobby is truly an exemplary accessible application that will work well across multiple operating systems.

Since Bobby's release, CAST has been employing user feedback to improve the program and provide information to the WAI and others interested in accessibility. Through continued field testing and the collection of user feedback,

CAST will inform the ongoing development of Web accessibility page authoring guidelines.

Bobby 3.2 is the most current release and version 4.0 is under development at the time of writing. Bobby 3.2 features enhancements to usability and functionality, including:

- improvements to the readability of reports
- error descriptions rewritten for a less technical audience
- availability with IBM Self Voicing Kit which allows for Bobby's interface to be read aloud by the computer (downloadable Bobby)
- some translations of Bobby into other languages.

Bobby 4.0 will include major technical changes that allow for additional features:

- more robust page-checking capability
- an API (Application Programming Interface) that allows technical users to write custom page-checking modules for internal use or wide distribution
- an API that allows third-party programs, such as page authoring tools and accessibility repair tools, to control Bobby and its results
- a single international version that allows you to choose a preferred language.

The way that pages are examined will be updated. While Bobby is currently able to check many of the *Web Content Guidelines*, there are many that it is not able to examine. Bobby's error reporting also does not align fully with the organization of the *Guidelines*. Action items to address this include the reorganization of Bobby's checks to match the WAI *Guidelines*. In co-ordination with members of the Web Accessibility Initiative, the programmatic techniques used in the analysis of complex checkpoints will be examined. Bobby's internal code will be rewritten to provide a more flexible view of the HTML page and enable the development of more robust checking logic. The text explaining page errors in Bobby reports will also be rewritten to be more user friendly.

CAST is partnering with several organizations to create non-English versions of Bobby. Translations that are currently under development include Dutch, French, German and Japanese. They will be made available as separate programs as they are completed. In the future Bobby will support on-the-fly switching to the language the user selects.

We must emphasize that Bobby is an access evaluation tool which flags accessibility violations and provides suggestions on how to make improvements. Bobby does not, however, have the capability of repairing the problems. In partnership with the Trace Research and Development Center and the University of Toronto Adaptive Technology Research Center, a new tool is being developed that will help fill this gap. The new tool will use the information from a Bobby report and walk the user through the process of repairing the page.

## 8.6 CONCLUSIONS

As Web site developers become aware of accessibility issues, barriers are coming down. However, this increased awareness raises important challenges for site designers, educators, government agencies, corporations and organizations as they now recognize how much more needs to be done to achieve universal access.

For instance, with the Web's rapid growth, an alarming number of new inaccessible sites are created each day. The assistive technologies used by individuals with disabilities cannot keep up with the Web's growth and that of new technologies that may present barriers.

The answer lies in universal design. Universal design is the concept that information systems, software products, Internet tools and learning models are designed at the outset so that they are usable by everyone, including those with disabilities.

The notion of a universally designed World Wide Web challenges society to think about plurality to consider all individuals, regardless of age, ability, race, or economic or cultural background when developing new technologies. The existing notion is that building in access is costly and only benefits a small portion of the population. In fact, the opposite is true. Creating universally designed products costs less than adding assistive supports later and universally designed technologies benefit a broader range of users.

While Bobby offers one way to create a more accessible World Wide Web, it brings to light significant challenges. CAST and other organizations are always looking for ways to promote universal design. However, our society must be prepared to meet these challenges as openly as it embraces new technologies.

The significance of the Bobby project will be revealed by the extent to which Web developers throughout the world are willing and able to detect and repair accessibility problems on existing Web sites. Success will also be measured by the impact that Bobby has on the development of accessible Web sites as they are being developed. Continuous feedback from those who use both the online and client applications indicates that most Web site developers who use Bobby for a brief period of time find that they create fewer barriers to accessibility. Bobby is then used to provide a check of the work rather than to expose problems for the first time.

We continue to hold the vision that the World Wide Web will be usable by everyone, including individuals with a wide range of disabilities. Removing the barriers to access will ensure that the Web will serve as an environment supporting recreation, education and work.

To best understand how Bobby works, go to www.cast.org/bobby/ Type in any URL and Bobby will quickly provide a detailed report of accessibility problems and show you how to correct them. Select a number of different browsers and Bobby will check the page's compatibility. Links on the Bobby home page also connect users to background information and documentation in the form of Frequently Asked Questions (FAQ). A link to the downloadable Bobby application FTP site is also available from the Bobby home page.

## ACKNOWLEDGEMENTS

The author wishes to thank Josh Krieger, the original Bobby developer under contract to CAST, for providing technical notes used in this chapter. The author appreciates the hard work and dedication of the W3C–WAI staff and working groups for providing *Web Content Accessibility Guidelines* upon which Bobby bases its accessibility evaluation. Michael Cooper deserves credit for his leadership in continuously striving to improve the Bobby service in his role as manager of the Bobby project at CAST. Finally, thanks to CAST's management and resource development office for supporting the ongoing development of this important international Web accessibility service.

# Part 4

# Existing Guidelines

## CHAPTER NINE

# Nordic Guidelines for Computer Accessibility

Clas Thorén

### 9.1 SCOPE AND OBJECTIVE

*Nordic Guidelines for Computer Accessibility* is published by Nordic Cooperation on Disability, which is an organisation under the Nordic Council of Ministers, i.e. the governments of Denmark, Finland, Iceland, Norway and Sweden. The current edition of this publication was issued in 1998 and replaced the same title issued by the former Nordic Committee on Disability in 1993.

The origin of *Nordic Guidelines for Computer Accessibility* goes back to the late 1980s when accessibility requirements were introduced in a procurement of PCs for use in Swedish government agencies. Inspired by the Swedish initiative, the Nordic Committee on Disability set up a project team of experts from Nordic institutes for technical aids, with the purpose of producing a guideline on accessibility to personal computers. The intention was to target the guideline to the ICT business, i.e. developers, manufacturers, purchasers and standardisation groups. The guideline was published in 1993 and updated in 1998 due to further technical developments. The updating was made with the support of a reference group of experts from within and outside the Nordic countries. The content was focused more, but not only, on purchasers and ICT strategists.

The scope of the *Nordic Guidelines for Computer Accessibility* is both wider and narrower than that of a design guideline.

It has a wider scope in the sense that it is targeted at not only designers, but also—and to a greater extent—procurers and standardisers. This being so, it should be recognised that procurements, in particular large-scale public procurements, are powerful instruments for influencing the market. The suppliers, and consequently the designers, are sensitive to specifications used by major economic players such as public procurers.

The scope is narrower in the fields of application software and Web pages, where a large number of design guidelines exist. Instead of elaborating additional competing guidelines in these fields, or making a condensed synthesis of the existing guidelines, *Nordic Guidelines for Computer Accessibility* gives a list of examples of such guidelines.

The design process, in particular the user-centred design process, is iterative. Approaches and prototypes are tested by panel groups and modified according to their feedback. A design guideline may therefore be considered as a list of required and recommended features, of which the designer is expected to take into account all the requirements and as many as possible of the recommendations. The designer and the customer may, however, agree to disregard or modify one or more requirements due to technical or economical constraints which may become obvious during the design process. A design guideline may express a long-term goal—an ideal situation that the designer is supposed to aim for, but implicitly recognise that the goal will not be achieved 100%.

The procurement process is different from the design process. Public procurements are strictly regulated and are definitely not iterative. The procurer is not allowed to add, reformulate or delete requirements and recommendations after the tenders are received. Moreover this is included in the code of good practice for any procurement. If a mandatory requirement is for some reason disregarded in the evaluation of tenders, the suppliers may have submitted their tenders—or refrained from submitting a tender—on false presumptions. Furthermore, since a call-for-tender is based on a decision to purchase, the requirements must be capable of being satisfied by a reasonable number of products existing in the market. Consequently, the requirements and recommendations to be included in a call-for-tender have to be considered very carefully. Once they are formulated and sent to the suppliers, they must not be changed. This is contrary to the design process, which needs to cater for a certain degree of freedom for the designer.

Since the *Nordic Guidelines for Computer Accessibility* is intended to be both a design guideline and a procurement guideline, attempts have been made to achieve a balance between:

- improvement of the accessibility of personal computers and other workstations, in order to meet the needs of disabled and elderly people
- procurers' and suppliers' needs for requirements that are easy to understand and evaluate
- the need to avoid distorting competition between suppliers.

## 9.2 STRUCTURE

*Nordic Guidelines for Computer Accessibility* is divided into two parts. Part I describes what is meant by accessible information and communication technologies (ICT) and gives the rationale for inclusion of accessibility requirements in ICT procurement, ICT standardisation and ICT design. Part II presents a set of functional requirements which meets the need for accessibility of personal computer systems operated by the end-user. Except for a section on the accessibility of Web pages, and the recognition that Microsoft Windows has a market-leading position with today's desktop and portable systems, the intention has been to keep this publication independent of applications, technologies and vendors.

*Nordic Guidelines for Computer Accessibility* defines accessibility as a set of properties that are built into the product, service or system from the outset, enabling people within the widest range of abilities and circumstances as is commercially practical to access and use it. This approach is often referred to as the principle of *Design for All* or *Universal Design* and should be taken into account not only by designers, but also by manufacturers, value adders, retailers, purchasers, strategists and end-users.

*Design for All* cannot, however, satisfy *all* needs of *all* people. There will always be people who need some kind of assistive devices, i.e. specifically designed add-on input and output devices that compensate for different kinds of disabilities. Such devices enable an individual independently to complete tasks that previously could not be performed without assistance, due to the disabling condition. *Nordic Guidelines for Computer Accessibility* requires any ICT system to be designed to enable the user to attach an appropriate assistive device, perform the same tasks and achieve the same results as any other user.

*Nordic Guidelines for Computer Accessibility* explains the benefits of accessible ICT. There are benefits for all users, for users with disabilities, for the individual ICT suppliers and for the industry at large.

## 9.3 TARGET GROUPS

*Nordic Guidelines for Computer Accessibility* addresses three main target groups:

a) One purpose is to provide *purchasers* with a reference document to be used in or referred to in the call-for-proposal, by providing criteria for the selection of the most accessible equipment. In addition, *IT managers* responsible for corporate policy-making and planning will be provided with a reference document for taking accessibility into consideration in their work.

b) A second target group is *standards bodies and organisations*. The design of information technology is to a great extent determined by specifications, made by standards bodies, organisations and vendor associations with the purpose of establishing and maintaining official and de facto standards. It is of great importance that accessibility requirements are seriously taken into account in this work. Ideally, accessibility should be a feature that is permeated throughout all standards (see Chapters 4 and 5).

c) A third purpose is to encourage *manufacturers and vendors* of computer hardware and software to apply the accessibility requirements in the developmental process.

*Nordic Guidelines for Computer Accessibility* defines a set of functional requirements meeting the needs for accessibility of ICT equipment that is operated by the end-user, e.g. personal computer, display, keyboard, mouse, printer, operating environment and application software. This includes various kinds of display and keyboard-based equipment, such as desktop personal computers, portable personal computers and high-performance technical workstations used in all sectors of society—work-life, public services, schools and home.

*Nordic Guidelines for Computer Accessibility* is intended to be usable as a reference document in procurement. For this reason, it is presupposed that the basic product qualities that are needed by all users are expressed in the call-for-proposal, either explicitly or by referring to corporate reference documents. Such corporate documents would normally be referring to regulations, standards or other reference documents which need to be followed.

Due to rapid technological development, no attempt has been made to classify the requirements into categories such as mandatory, recommended or desirable. The verb 'should' is used throughout the requirements.

## 9.4 INTERNAL AND EXTERNAL USE

*Nordic Guidelines for Computer Accessibility* points out that the accessibility requirements are somewhat different for ICT systems intended for the public and corporate internal systems. For ICT systems intended for the public, such as an information kiosk or a terminal at a public library, the user is unknown and can be anyone. A public terminal can be adapted to the individual user only to the extent permitted by the system itself. There will be no possibility for a specific user to attach an assistive device other than by, for example, plugging an earphone in a standard socket intended for the public user. Therefore, the accessibility of such a system needs to be maximised in order to allow the widest range of consumers or citizens to use the system (See Chapter 14).

Accessibility of internal corporate systems means accessibility for employees. Any current or future employee may be or become affected by circumstances that makes the access to and use of ICT products, services and systems difficult. The employer should ensure that systems are purchased which provide both basic accessibility and adaptability to individual cases where necessary, i.e. the system should allow modifications, selection of a different options, the connection of an add-on assistive device or alternative equipment.

## 9.5 CONTENT

*Nordic Guidelines for Computer Accessibility* advocates the application of the principle of *open systems*, i.e. hardware and software complying with international and de facto standards. Open systems give the user freedom of choice of products and suppliers, with preserved system structure, allowing applications to live on despite rapid technical development. Assistive devices—hardware and software—should also comply with the principles of open systems, whereby a person who needs a particular assistive device will be able to use the application during its lifetime and also exchange the device if the impairment changes.

Part II of *Nordic Guidelines for Computer Accessibility* contains a set of accessibility requirements subdivided into requirements on the components of a personal computer system:

- The *technical platform* should be open and allow generally available application software to be accessible and assistive hardware and software to be installed and executed. In practice, the platform is in most cases equivalent with Windows. However, the Guidelines are intended to cover other platforms such as Apple Macintosh and UNIX-based systems; hence a number of basic accessibility requirements are given, such as a customisable user interface, keyboard access to all features, selectable output modes, etc.

- The requirements on the *system unit* are mainly related to its capability of connecting assistive devices.
- The *controls* of the hardware, such as turn knobs, switches and thumb wheels, must be easy to find, reach, identify and operate. The setting of hardware features are often software controlled today but for the sake of generality a comprehensive set of accessibility requirements are provided.
- *Alarms, warnings, status messages,* etc. should be able to be presented in selectable modes—auditory, visual or tactile.
- The accessibility of the *keyboard* is a crucial feature of a computer system. Voice recognition systems are entering the market, but for languages other than English the key input will be the main input mode for text during the foreseeable future. Furthermore, the keyboard is essential for people who have difficulties in operating a mouse or other pointing devices. The Guidelines cover requirements on the finding, identification, hitting and pressing of keys.
- A number of accessibility requirements of *pointing devices* are provided.
- Most accessibility problems associated with the *display* are software-related and are mostly solved by setting parameters or taking advantage of accessibility features built into the operating system. For people with low vision, the basic ergonomic requirements on displays are important; hence examples of such requirements are provided.
- A section is devoted to the *printer*, although requirements on controls, alarms, etc. are covered by other sections.
- People wearing hearing aids are concerned about *electromagnetic features* and the *noise* emitted by the equipment. Requirements on these features are provided.
- For *application software*, a number of design guidelines are available which provide detailed recommendations, advice and explanatory information on how different system features could be designed in order to maximise accessibility. Examples on software accessibility guidelines are provided.[1] For further information, the reader is referred to the Web site of the INCLUDE project—www.stakes.fi/include (see Chapter 13).

The Guidelines discuss how to include accessibility requirements in procurements of software. There are two aspects of this issue. Firstly, standard software such as word processors, Web browsers, spreadsheets, etc., are, in fact, seldom procured by open calls for proposals including detailed requirements on functions and performance. Instead, they are included in a package combined with the computer or determined in a corporate strategy. Secondly, the business requirements on specific software such as CAD/CAM, payroll systems, etc. may be satisfied by products available on the market. If not, a new software program will

[1] *Macintosh Human Interface Guidelines*, 1992; Vanderheiden, 1992a and 1992b; Bergman and Johnson, 1995; *Microsoft® Windows® Guidelines for Accessible Software Design; Common Desktop Environment: Style Guide and Certification Checklist* (Hewlett-Packard, IBM, Novell and Sun Microsystems).

be developed to be tailored to the needs of the customer. In this case, the developer should be requested to apply a software accessibility guideline. The Guidelines provide approaches for covering the accessibility issues in procurements and the design of application software.

- The same approach as for the design of application software is applicable to the design of *Web pages* and the designer is recommended to apply a Web design guideline.[2] The most prominent is the one provided by the Web Accessibility Initiative, a project associated with the World Wide Web Consortium and available at www.w3.org/wai (see Chapter 10).
- *Documentation* and online help text are necessary to make full use of the versatility of modern standard software. Requirements on help text and other documentation is provided.
- The log-in system often includes an *authorisation system*, where the technical solution may make it difficult or impossible for some users to enter the system. Requirements of the security systems are provided.

## 9.6 SUMMARY

The *Nordic Guidelines for Computer Accessibility* has been widely referred to and utilised in other guidelines and informative documents, e.g., ISO TS 16071, *Guidance on Accessibility of Human–Computer Interfaces*. It has been translated into Japanese and Italian and is used in Sweden and Iceland in public procurements of PC equipment. Other examples of use are as an element in training courses in the former COMETT programme and in the Open University, UK. Furthermore, it is the basis of the ACCENT project (*Accessibility in ICT Procurement*), the results of which are guidelines for including accessibility in the different phases of a public ICT procurement process. ACCENT is financed by the EU initiative SPRITE–S2 and the Nordic Development Centre for Rehabilitation Technology.

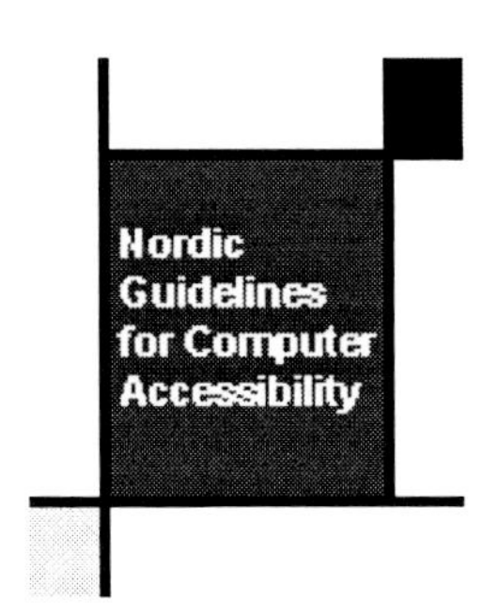


[2] Examples given are: CCTA, *Writing for the Web*, (UK: CCTA), www.open.gov.uk/webguide/wgindex.htm; US Department of Education World Wide Web (WWW) Server Standards and Guidelines, http://inet.ed.gov/~kstubbs/wwwstds.html; *Sun on the Net: Guide to Web Style*, www.sun.com/styleguide/

*Nordic Guidelines for Computer Accessibility* can be ordered in printed form, free of charge, from The Nordic Cooperation on Disability, email: nsh@nsh.se

CHAPTER TEN

# Guidelines for Web Accessibility

Jan J. Engelen

## 10.1 ACCESSIBILITY AND USABILITY

*Accessibility* is a central concept in the literature concerning technology for disabled people, but an even more important item is *usability*.

It is important to make the accessibility/usability distinction very clear. According to dictionary definitions, something is accessible if it can be reached. Accessibility can be defined as a set of properties that are built into the product, service or system from the start, enabling people within the widest range of abilities and circumstances, as is commercially practical, to access and use it.

Let us take an example from the computer field. A word-processing package might be accessible if the user *can* perform the functions contained within it. Usability implies consideration of the ease with which the user can perform those functions, how effectively and efficiently (in terms of needed time and avoidance of repetitive tasks) the user can write a paper with the word processor.

Docx *et al.* (1998) made a distinction between physical, technical, structural, cognitive and enjoyable access. From these considerations, it becomes clear that most users will appreciate usability more than accessibility.

This is also true for access to the World Wide Web. It is almost always possible to find out what is on a Web page if the disabled user has adapted and sophisticated equipment at hand and spends much time using it. But for the Web information to become usable to disabled users, some guidelines must be followed by people drafting and editing the Web pages.

Later we will discuss extensively the *Web Accessibility Initiative (WAI) Accessibility Guidelines*, taking into account that, in my opinion, what the WAI actually wants to achieve is the usability of the Web pages.

## 10.2 CURRENT HARDWARE AND SOFTWARE FOR WEB ACCESS TO PRINT-HANDICAPPED USERS

Web pages are developed for graphical user interfaces (Windows, UNIX X-terminal, Macintosh). These interfaces are particularly difficult for the blind. Nevertheless, several technical solutions to Web browsing are available for visually impaired persons.[1]

a) For the users of *standard browsers* (NETSCAPE NAVIGATOR, INTERNET EXPLORER, etc.), *screen reader* software is available. Screen readers are

[1] More details can be found at: www.w3.org/WAI/References/Browsing

programs running in the background on a computer that are able to translate the information of a foreground application (e.g., a word processor or browser) into large characters on the screen, or to output codes driving a speech synthesiser or a Braille reading line. Screen readers with good functionality are now on the market and there is considerable further development underway, especially as in recent years the makers of operating systems have collaborated more than in the past. Through these developments, it is possible for blind and partially sighted people to have usable access to a standard graphical WWW browser.

Recently several *add-on software packages* became available that support the users of standard browsers by giving extra information (about links, graphical objects, table layout, forms, etc.) in supplementary windows (e.g., SIMPLY WEB, WEBWIZARD, etc.). Experimental solutions such as HELPDB, developed by Mark Novak at the Trace Center,[2] also fit into this category. The creators of the OPERA Web browser, on the other hand, have included many accessibility features in their standard product.

b) Several *specialised browser* packages, usable by visually impaired persons or developed specifically for this user group, have appeared on the market. They can be divided into two classes:

- specially developed software (e.g., PWEBSPEAK, LYNX, NETTAMER)
- new accessible user interfaces to existing browsers (e.g., the BRAILLESURFbrowser, working in conjunction with Microsoft's Internet Explorer; IBM's HOME PAGE READER, presenting an accessible human–computer interface to Netscape Navigator; the Oxford Brookes talking browser, also based on Internet Explorer).

c) Finally, several browsers have been developed, based on *design for all* techniques. Their human interface changes according to the needs and the possibilities of the user. Especially noteworthy is the AVANTI[3] browser, in which the feasibility of employing adaptability and adaptivity techniques in rendering Web-based, multimedia information systems accessible to people with special needs is demonstrated.

d) Unfortunately, there are a number of Web features which are very difficult to make accessible and for which in general one will have to rely on text descriptions. Included in that group are video sequences (MPG), Shockwave animations, Java applets, ECMA script, etc.

Adobe's Portable Document Format (PDF) is a special case: although documents in PDF are generally inaccessible, several solutions have been put forward ranging from the use of intermediate document transformation computers (proxies) to text extraction via Adobe's Web site or via their mailserver. Adobe also provides a special module as an add-on to Adobe Acrobat Reader, the standard PDF reading software.[4]

---

[2] Downloadable from: www.dinf.org/csun_99/session0087.html

[3] AVANTI is the name of a European project on advanced communication and telecommunications (ACTS): www.avanti-acts.org/

[4] Details from: http://access.adobe.com

## 10.3 STRUCTURED VERSUS UNSTRUCTURED ELECTRONIC DOCUMENTS

Long before the WWW was invented, its language took form. In the second half of the eighties, a special method for adding structure to an electronic document was standardised by the International Standards Organisation (ISO) under the name Standard Generalized Mark-up Language (SGML). From then on, there was a system (although, even nowadays, still considered too complicated by many) to make the meaning of parts of a document clear and this independently from such ephemeral things as company standards (a whole series of Microsoft word formats, a somewhat smaller series of, for example, WordPerfect formats).

In a special support programme of the European Union called Technology and Blindness, attention was paid (among several other issues concerning visually impaired persons) to the issue of electronic newspapers for the blind. In a 1990 meeting in Helsinki, a system to add some tagged information to otherwise plain ASCII text was developed.

In 1991, the European TIDE[5] project CAPS (Communication and Access for Persons with Special Needs) was set up. The CAPS work has demonstrated the importance of structure in electronic documents. The normally sighted user obtains a significant amount of information from the layout of a document—titles in bold, bulleted indents, emphasised sections in italics. These are crucial when browsing through a long document. In most electronic documents the structure is not accessible directly, being defined primarily to create the visual end product. To make this information available in order to aid the print-disabled user to browse—or navigate—within a document, the structure needs to be defined explicitly within the electronic document. The CAPS work has deepened the conviction that the technological possibilities created by standardised structured electronic documents (using SGML) can now form the basis of a generic model for dramatically improving the access to information for blind and partially sighted people.

CAPS collaborated intensively with the late Yuri Rubinsky, founder of the SoftQuad SGML company, who used his personal charisma to convince mainstream SGML (and later HTML) companies to pay attention to accessibility.

In 1989 the idea of a global hypertext space in which any network-accessible information could be referred to by a single 'Universal Document Identifier' was put forward by Tim Berners-Lee, then working at CERN (the European Particle Physics Laboratory in Geneva). In 1990 he wrote a program called 'WorlDwidEweb', a point and click hypertext editor which ran on a 'NeXT' machine. In 1991 all ideas and software (one server and two browsers) were made available to the Internet community, still very much research-oriented at that time.

The World Wide Web was born, but it would take several years more before it really caught on.

It was very fortunate that WWW functionality required the existence of a structured document language as well. HTML (HyperText Mark-up Language), version 1, was developed. Technically it is a simple SGML DTD—a Document

---

[5] The European Commission has supported research on assistive technology in various programmes. TIDE (Technology Initiative for Disabled and Elderly People) was the first one in this series (1991).

Type Definition—permitting the fast and cheap construction of the first WWW browsers.

History has meanwhile proven that HTML needed to be more versatile and several HTML versions appeared (the popular ones are versions 2, 3, 3.2 and 4). The Extensible Mark-up Language (XML), which combines much of the rigour of SGML with the essential simplicity of HTML, has recently emerged as a candidate for structured document interchange on the Web.

## 10.4 THE NEED FOR GUIDELINES

Many organisations and researchers working with and for visually impaired persons found out that even a relatively simple language such as HTML allowed the creation of documents that present severe difficulties to reading-impaired users. At the same time, the World Wide Web was considered as a key enabling technology for increasing access to information for blind and partially sighted people. The Web is of more value to them, as quite often they do not have other possibilities to access the information.

### 10.4.1 TRACE

The Trace Research & Development Center is a part of the College of Engineering, University of Wisconsin–Madison. Founded in 1971, Trace has been a pioneer in the field of technology and disability (see Chapter 18 and Trace.wisc.edu). It has been continuously working on ways to make standard information technologies and telecommunications systems more accessible and usable by people with disabilities. Their work is primarily funded by the National Institute on Disability and Rehabilitation Research (NIDRR) (US Department of Education).

Historically, one of the first overall attempts to develop guidelines for authors of accessible Web pages were collected at the Trace Center. Their comprehensive list *The Trace Unified Web Guidelines (version 8)* (Vanderheiden *et al.*, 1997) constitutes the last version produced at Trace and became the starting point for the WAI guidelines (see below).

### 10.4.2 ICADD

In 1991 an international co-operative effort led to the ICADD initiative—the International Committee on Accessible Document Design. Ideas on how documents should be structured in order to be usable were exchanged and a (short—22 tag) ICADD DTD (i.e., a structure description document) was produced. This was during pre-WWW time and ICADD's major objective was to develop a *simple* system so that publishers could be forced to deliver electronic texts of school books to a centralised repository service (as already was the case in Texas in 1992!). Later, several ICADD proponents continued their actions within the WAI initiative (see below).

### 10.4.3 HARMONY

In Europe several groups also studied the possibilities of drawing up guidelines for usable Web pages. The consortium of European organisations that undertook the HARMONY[6] project, a support action within Europe's TIDE (Technology Initiative for Disabled and Elderly People) Programme, had done extensive user testing of Web guidelines at the Royal National Institute for the Blind (Peterborough, UK). The HARMONY consortium came up with a basic list of guidelines promoting the principle that applying a dozen rules could guarantee that 90% of the Web pages become accessible.

The HARMONY consortium also organised the specialist workshop *Reading Without Paper* (Bruges, Belgium, July 1996), where Internet access by visually handicapped persons was extensively discussed.[7]

### 10.4.4 W3C–WAI

#### *10.4.4.1 History*

In the middle of the nineties a new organisation, the World Wide Web Consortium (soon to be known as W3C) was set up by a large group of interested organisations and companies in order to study the WWW evolution as a whole and to develop vendor-neutral recommendations on all rapidly evolving, technical aspects of the Web. W3C became in a short time the *de facto* 'organiser' of the WWW. Tim Berners-Lee, inventor of the Web (see above), became the director of W3C. At the beginning of the year 2000, the W3C had over 400 member organisations.

Since its inception, the W3C has had an official activity area devoted to accessibility for people with disabilities, thanks to the continuous efforts of Mike Paciello of the Yuri Rubinski Insight Foundation and Georges Kerscher, who were also founding members of ICADD. After a meeting with high officials at the White House in January 1997, the World Wide Web Consortium officially launched the Web Accessibility Initiative (WAI) with the formal support of US President Clinton on 6 April 1997.

#### *10.4.4.2 Objectives*

When the WAI International Project Office opened, it was stated that W3C was to take on three roles with respect to accessibility:

1. Act as a central point for setting accessibility goals for the Web: This requires W3C to collaborate with external organisations that represent people with disabilities to generate a widely accepted set of goals and guidelines that take into account the needs of the user community, the details of the technology, and engineering realities. W3C already fills this role in several other areas of technology and it is a logical extension to a new user community.

---

[6] www.esat.kuleuven.ac.be/teo/harmony

[7] Proceedings available: mailto:info@sensotec.be

2. Act as an advocate for people with disabilities to the Web development community: As the internationally acknowledged organisation and leader for World Wide Web development, the W3C acknowledges its responsibility for advocating Web accessibility for people with disabilities. As the Web user interface and infrastructure continues to evolve, the W3C will work to help its members become proactive in their efforts to design and develop the Web in a way that considers the user needs of people with disabilities.
3. Act as an advocate for people with disabilities to the Web content community: W3C already serves as a neutral party for distributing information about Web technology. W3C would like to extend this role to be proactive in explaining to content producers how best to use the technology to serve the needs of people with disabilities.

*10.4.4.3 Guidelines*

The WAI organised its actions according to World Wide Web Consortium practice—new items are discussed for a certain time by internationally established working groups, keeping in touch via e-mail. Most of the discussion groups also welcome comments by interested specialists. When a consensus is reached a draft of a 'recommendation' is made to be voted upon by W3C members.

W3C–WAI has established three W3C recommendations to improve the accessibility of the Web. They[8] are:

- The *Web Content Accessibility Guidelines*, released in May 1999, which explain how to make Web sites sufficiently accessible so that people with disabilities can use them with today's technologies. These guidelines were released with the support of industry leaders, as well as support from disability organisations internationally, access-research organisations, and governments interested in ensuring that information on the Web is accessible.[9]
- The *Authoring Tool Accessibility Guidelines*, released in February 2000, which provide guidance to developers of software used to build Web sites, so that the software will automatically ensure accessibility of much of the code used on Web sites. These guidelines were also released with support from industry leaders and from many organisations.
- The *User Agent Accessibility Guidelines* (March 2000), which explain how to make browsers and multimedia players more accessible and how to make them work better with some of the assistive technology that people with disabilities use.

---

[8] Information on W3C's Web Accessibility Initiative is available at www.w3.org/WAI, including links to the *Web Content Accessibility Guidelines* (www.w3.org/TR/WCAG), *the Authoring Tool Accessibility Guidelines* (www.w3.org/TR/ATAG) and the *User Agent Accessibility Guidelines* (www.w3.org/TR/UAAG).

[9] When the WAI was formed in 1997, there were already over 40 documents that had been written to address Web accessibility. The WAI-GL working group adopted the Trace Unified Web Guidelines (version 8) as the basis for their Web content guidelines document.

### 10.4.5 Europe and the WAI

In order to promote the work on the WAI guidelines also within the European context, two European WAI projects have been set up in recent years.

#### *10.4.5.1 WAI support action (TAP-DE 3467)*

The TIDE WAI project was set up to formalise the co-operation between the European Union and the W3C. The project was investigating, among other things, promoting the activities of the WAI International Program Office in Europe and how the EU policy rule to give people with disabilities equal access to electronic information could be implemented.[10]

#### *10.4.5.2 WAI-DA project (IST-1999-13470)*

The Web Accessibility Initiative—Design for All project started at the end of 1999 with support from the Fifth Framework Programme for Research and Development of the European Union. It is a follow-up of the previous project.

## 10.5 EVALUATION OF ACCESSIBILITY GUIDELINES

### 10.5.1 WAI checklists

Each of the WAI guidelines includes a simplified, prioritised accessibility checkpoint list, intended to give developers an easier means of following the guidelines by checking off the items that relate to their application or content.

- *The Web Content Guidelines Checklist*[11]
- *The User Agent Guidelines Checklist*[12]
- *The Authoring Tool Accessibility Guidelines Checklist*[13]

### 10.5.2 Self Evaluation Test

The Self Evaluation Test,[14] produced by the Public Service Commission of Canada, enables site designers manually to go through a number of statements to discover if their Web pages are likely to be accessible, with some fairly general suggestions of

[10] TIDE WAI was co-ordinated by INRIA (French Institute for Research in Computer Science and Control). Contact: Dr. Daniel Dardailler, of INRIA Sophia-Antipolis, e-mail: danield@w3.org The project description is located at: www2.echo.lu/telematics/disabl/wai.html

[11] www.w3.org/TR/WAI-WEBCONTENT/full-checklist.html

[12] www.w3.org/TR/WAI-USERAGENT/full-checklist.html

[13] www.w3.org/TR/2000/REC-ATAG10-20000203/atag10-chklist.html

[14] The WPASET (*Web Page Accessibility Self-Evaluation Test*) can be found at: www.psc-cfp.gc.ca/dmd/access/welcome1.htm

how to improve overall accessibility. This method is rather subjective but does have the advantage of directly involving the designers and actively encouraging them to adopt an accessible approach to all Web pages.

### 10.5.3 Evaluation of the guideline concept

Ergonomic guidelines have been produced in the past for many different fields. Recently an attempt was made to see if the WAI guidelines would fit into a generalised structure of accessibility guidelines (Wünschmann, 2000). A clear distinction could be made between the structure of ISO standards and the W3C–WAI recommendation, the latter being clearly less rigorous in its definitions and wordings.

### 10.5.4 Web accessibility symbols

In order to make clear to the readers that a Webmaster has paid attention to accessibility, a special symbol can be put onto the page. Sometimes there is a verification of the accessibility—sometimes the use of the symbol is voluntary.

Bobby[15] is a Web service designed to help Web site designers and graphic artists make their Web pages accessible to the largest number of people (see Chapter 8). It will help find design problems which prevent a Web page from being displayed correctly on different Web Browsers (or 'User Agents') without having to individually test the page on each of those programs. If a site passes the examination, it may carry the 'Bobby Approved' symbol.

The Web Accessibility Symbol that has been adopted and promoted by NCAM[16] is also recognised as the symbol for Web sites which are accessible. The use of the symbol is free and voluntary, and therefore there is actually no guarantee of accessibility associated with it. The Web Accessibility Symbol must always be accompanied by the Description (or 'D-tag'):

*A globe, marked with a grid, tilts at an angle. A keyhole is cut into its surface*

and the Alternative Description Text (or 'alt-text tag'):

*Web Access Symbol (for people with disabilities).*

Despite this, several other symbols are in use (e.g., 'Speech Friendly Site', 'Lynx Optimised') and can be found regularly on (hopefully, accessible) Web pages.

---

[15] The home page for Bobby is at: www.cast.org/bobby/

[16] NCAM is the US based National Center for Accessible Media. More information on the symbol can be found at: www.wgbh.org/wgbh/pages/ncam/webaccess/symbolwinner.html

## 10.6 THE GLOBAL ADOPTION OF WEB ACCESSIBILITY GUIDELINES: STATUS

### 10.6.1 Introduction

The page author guidelines of the W3C–WAI are meanwhile accepted by many organisations as the point of reference in these matters—they were initially based on existing (mainly Trace) documents, have been extensively commented on internationally and have been accepted as a W3C recommendation on 5 May 1999. Meanwhile translations in many languages have been made in order to diffuse the ideas worldwide.

Special information material was also produced by W3C–WAI for dissemination purposes,[17] such as a credit card-sized guideline overview, several PowerPoint presentations, a curriculum and a video.

Many organisations started promoting the WAI ideas all over the world. One of these was the COST219bis action (see Chapter 13), an international collaborative group whose main objective is to increase the availability of telecommunication services and equipment so that they are accessible to disabled and elderly people.[18]

### 10.6.2 USA

The US Access Board[19] is an independent federal (governmental) group which oversees the production of guidelines on accessibility for compliance with various legislative measures. The Access Board, in producing a set of guidelines required by a legislative mandate, will generally follow a process along the following lines:

- first convene a set of experts from academia, industry and consumer groups (an 'advisory committee') to produce a report making recommendations for guidelines
- produce a *Notice of Proposed Rule Making* (NPRM) document, which is then available for public comment
- After the public comment period, a final set of guidelines is produced.

The US Access Board played a major role in:

- The *Telecommunications Act Accessibility Guidelines* (TAAG). These guidelines, also referred to as 'Section 255', set out requirements that telecommunications equipment manufacturers and service providers make their

---

[17] Details from: www.w3.org/WAI

[18] An overview brochure *Telecommunications—Guidelines for Accessibility* and a guidebook *Producing Web Pages that Everyone can Access*, were produced by COST219bis members and distributed at Telecom99—worldwide the largest exhibition fair on telecommunications. Both are available from: www.stakes.fi/cost219

[19] www.access-board.gov/

products accessible—*where readily achievable*. These guidelines have been adopted by the Federal Communications Commission (FCC).

- The *Rehabilitation Act 'Section 508' on Information Technology Guidelines*. This Act states that where it is not an '*undue burden*' on federal agencies, they must favour purchasing information technology products that are accessible to people with disabilities. The aim is for manufacturers to compete with each other by providing accessibility to win federal contracts. On 7 August 2000, a revised amendment to the Rehabilitation Act of 1973 became effective. It clearly states that when federal agencies provide information to the public, information must be as accessible to people with disabilities as it is to members of the public who do not have disabilities. There is also the same requirement for agencies to provide access to information technology for employees with disabilities.

However, the mere existence of these legal items does not automatically guarantee widespread adherence to WAI guidelines.[20]

### 10.6.3 Europe

A long time passed before political Europe started paying attention to Web accessibility, although the TAP–WAI project (see above) had been funded by the European Union's Directorate XIII.

On a national basis the UK[21] and Portugal[22] especially have been able to make 'respect for accessibility' a requirement for governmental Web sites. The Flemish government also made sure that their tele-administration projects were aware of potential accessibility problems.

A resolution of the European Council in December 1996 endorsed the UN *Standard Rules on the Equalization of Opportunities for Persons with Disabilities*.[23] UN Rule 5 on accessibility demands access to the physical environment as well as access to information and communication. UN Rule 4 requests the full utilisation of high technology to improve the standard and effectiveness of assistive devices and equipment.

In 1998 the European Commission ordered a working document on *The Role of Telecommunication Systems for Elderly and Those with Special Needs—Ensuring Access for All*, in which among other items (such as a thoroughly developed demographic background) a call for political action was formulated.[24]

---

[20] See the statement of Judy Brewer, director of the Web Accessibility Initiative International Program Office before the US House of Representatives on 9 February 2000 (www.house.gov/judiciary/brew0209.htm)

[21] www.rnib.org.uk/digital

[22] www.egroups.com/message/eeurope-pwd/36?

[23] The *Standard Rules on the Equalization of Opportunities for Persons with Disabilities*, adopted by the United Nations General Assembly, 20 December 1993 Resolution 48/96, United Nations, United Nations Department of Public Information, New York, DPI/1454, April 1994. Available from: www.dinf.org/un_dinf/un_4896.htm

[24] The report was made by Evangelische Stiftung Volmarstein (Germany) under EU contract 48442.

Recently a few EU activities brought Web accessibility into higher gear.

#### *10.6.3.1 The eEurope initiative*

On 8 December 1999 the European Commission launched an initiative entitled *eEurope—An Information Society for All*, which proposes ambitious targets to bring the benefits of the information society within reach of all Europeans.[25] The initiative focuses on ten priority areas, from education to transport and from healthcare to people with disabilities. Action line nr 7 dealt explicitly with the fact that information and communication technologies (ICT) should be used to improve the quality of life and job chances of people with disabilities.

The eEurope initiative then underwent critical reviews. For action line nr 7, Mr Francisco Godinho, adviser to the Portuguese Minister of Science and Technology, organised an e-mail discussion forum via the Internet. This forum had 92 members, comprising 20 countries (Portugal, Spain, France, Belgium, Germany, The Netherlands, Italy, Austria, Ireland, Greece, United Kingdom, Sweden, Denmark, Finland, Norway, Brazil, Argentina, United States, Australia and South Africa). Members of the European Commission also participated in the discussions.

A special note (called *suggestion IC*) for the members of the European Council (i.e., the Prime Ministers of the European Union) was prepared in which a pan-European plan was suggested in order to have all public Web sites and those set up with public money (from the member states or from the European Union itself) made accessible.

This was confirmed in the eEurope action plan of May 2000,[26] where the following item is explicitly mentioned:

| action | by | deadline |
|---|---|---|
| Adoption of the Web Accessibility Initiative (WAI) guidelines for public Web sites. | European Commission, Member States | end 2001 |

Although the final content of the eEurope initiative is still under discussion, some groups think that eEurope is already a first step in the planning of the sixth European Framework Programme for Research.

#### *10.6.3.2 The Commission's Communication of May 2000: "The European Commission acts to improve the access for the disabled at the workplace"*

On 15 May 2000, EU Commissioner Anna Diamantopoulou published a communication[27] outlining the means by which the European Union should improve access in every sense for the disabled. As part of the item *Addressing the Digital Divide,* it is suggested that:

---

[25] europa.eu.int/comm/information_society/eeurope/index_en.htm

[26] More about eEurope from: europa.eu.int/comm/information_society/eeurope/pdf/actionplan_en.pdf and europa.eu.int/comm/information_society/eeurope/actionplan/actline2c_en.htm

[27] The text can be found at:
europa.eu.int/rapid/start/cgi/guesten.ksh?p_action.gettxt=gt&doc=IP/00/477|0|RAPID&lg=EN

"all the European Institutions and the Member States endorse the existing Web Accessibility Initiative guidelines, making the design and content of all public Web sites accessible to people with disability" and

"the Commission requests European normalisation bodies to determine, in close co-operation with organisations representing the needs of elderly and disabled persons, ... requirements for standardisation to ensure accessibility for disabled and elderly people within the information society."

## 10.7 CONCLUSIONS

Research on document accessibility started at the same time as the documents described in this chapter became available. Likewise research on Web accessibility progressed in parallel with the development of the Web itself.

After some ten years of research and lobbying work, we now are in a phase where political take-up of the ideas on accessibility and the rights of reading-impaired users to be provided with usable documents have been recognised. This has been made possible only through truly international collaboration (Web Accessibility Initiative), using the Web itself. All national initiatives as described above have been brought together within the WAI cradle and accessibility research became an international issue at the end of the last millennium.

The examples of take-up in the US and in several European countries have now led to the adoption of accessibility rights by the European Commission itself (eEurope 2002 initiative).

However, in this rapidly changing domain many matters remain unresolved and will need to be investigated in the years to come. Accessibility of XML documents, of mobile access devices and of Internet on miniature hand-held computers still must be defined and tested.

A whole new class of electronic documents is also bound to appear on the market (and already has done to some degree): electronic books, readable on portable pocket-book-sized screens. What about their usability for reading-impaired persons?

From this chapter it is clear that accessibility more and more can be considered as a 'never ending story'—just as in other ICT developments.

CHAPTER ELEVEN

# USER*fit*: User Centred Design in Assistive Technology

David Poulson and Neil Waddell

## 11.1 INTRODUCTION

As a discipline, assistive technology has grown out of rehabilitation and medical engineering, the emphasis being placed on the development of products which can enhance or replace lost body function. Products have often been developed by medical experts, with little emphasis on understanding the wider needs of the products' users, and taking into account broader issues than the clinical value of the product.

Design as a discipline has changed significantly over the years, with a move towards greater user participation and the involvement of multi-disciplinary design teams. In many respects design in assistive technology has been left behind in the wake of this evolutionary development, but this is hardly surprising given that products in this area are designed with the medical needs of a particular client being paramount. Often products need to be extensively modified to the needs of an individual client, which brings to the sector the attributes of a craft industry rather than the mass production associated with modern industrial design. The emphasis on function is necessary in assistive technology, but products also need to be constructed based on a broad understanding of potential users' needs, rather than solely on the opinions of medical experts whose concerns are primarily with the clinical condition of the client. Methods are needed which allow end-users of such technology to make some contribution to their development, and which address broader issues such as quality of life and acceptance of technology.

It is also important to remember that the appearance of a product communicates much more than mere functional capability. In today's consumer society the styling of a product is a major marketing issue and is often, consciously or not, why the buyer will select one product over another. Products therefore say something about us, about our personality, our social status, our needs and our capabilities. This can be either positive or negative. In the case of assistive technology, where a product has been designed to aid a person with a disability, this is almost always negative and can result in stigmatisation. This in turn can further disable the individual in that by broadcasting that message to the outside world, the individual is considered a less than capable member of society. A challenge exists, therefore, to blend functionalism with dignity.

Surveys of European Commission (EC) funded projects carried out by the USER (Usability Requirements Elaboration for Rehabilitation Technology) project in 1992 emphasised that developers in the telematic community often lacked the skills to take usability issues into account (Poulson and Richardson, 1998). As telecommunications systems and advanced technologies (like speech recognition, robotics and virtual reality) become more prevalent in the assistive technology sector, usability issues become even more important. Some form of structured design approach and the need to document becomes especially important when the system being developed is a complex one and when the design team is large and multidisciplinary. It is then that good communication becomes vital.

## 11.2 EMBRACING A USER-CENTRED PERSPECTIVE TO ASSISTIVE TECHNOLOGY

An important point to note is that disability is situation-specific. Disability is not an attribute of the individual—it is the product of the interaction between the individual and their environment. For example, someone who uses a wheelchair may be 'disabled' amongst a group of individuals climbing a mountain but perfectly 'able' amongst that same group of individuals sitting round a table having a discussion. For someone with profound hearing loss the situation may be totally the reverse. Assistive technology devices are designed to compensate for loss of functional ability and therefore tend to fall into the category of medical aids rather than desirable products. One way to improve on this position, and begin to remove the stigma associated with using disability aids, is to view the whole area of product design from a much wider perspective. There are essentially three ways to assist individuals with disabilities:

1. Change the individual, e.g., via medical intervention or intensive and specialised education
2. Provide the individual with assistive devices to enable them to adapt better to the environment
3. Change the environment.

Options 1 and 2 are familiar to us and represent much of what is done presently and has been done in the past. Option 3 is more challenging but is increasingly regarded as potentially the most efficacious approach. *Universal Design*, *Barrier-Free Design*, *Design for All* and *Inclusive Design* are all terms used to describe an approach to product design that aims to widen the potential user population, to include as many people as possible with a wide range of abilities, and in a wide range of environments.

## 11.3 USER-CENTRED METHODOLOGIES IN ASSISTIVE TECHNOLOGY

There has been a significant move towards a more customer-oriented approach to design in European rehabilitation and the assistive technology sector, but there is some indication that this approach is more mature in the United States compared to Europe. For example, the RERC–TET Center for Assistive Technology (Jain and Usiak, 1997) adopts a problem-oriented approach to project development which focuses development activities on user needs rather than technology. A key aspect of this approach is the use of focus groups held with the primary stakeholders (end-users, care providers and prescribers). These are used both in the initial problem definition phases of development, and in later prototype testing and evaluation activities. The value of this approach has been demonstrated for a number of systems, with the Center effectively acting as an advocacy agency between end-users and developers.

A customer needs-oriented approach is also being used at the University of Texas (Tran *et al.*, 1997; Wieck *et al.*, 1997), following a design methodology described by Ulrich and Eppinger (1995). This approach provides a set of structured checklists for design, covering the phases of customer needs. These are then matched against engineering specifications. In both cases, emphasis is given to a problem-centred approach to development, and on matching user needs to specification using design checklists and matrices of product requirements matched against intended functionality.

In addition to a more user-oriented approach to product specification, there has been a shift towards a greater degree of user involvement in assessment activities. For example, a systems-oriented approach to evaluation has been described by Bain (1997), who argues that consumers' abilities, goals and needs should be considered in design, along with users' goals and tasks, and the environment in which they operate. Bain also provides a systematic approach to development and evaluation based on the close user involvement of potential users in design activities.

Morgan (1993) has also described the importance of adequate assessment with the use of new technology and notes that many people who are disabled have purchased inappropriate equipment due to a lack of suitable advice. Morgan also believes that it is important to assess individual needs and to give the individual concerned the choice as to whether they use technology or not. The importance of training in use and follow-up support with the use of technical aids is also stressed.

There is also some evidence to support the observation that many people who have a disability may be using aids that are not the most suitable for them and in the past everyday living aids were often bought by end-users without any expert input (Stowe *et al.*, 1988). A recent UK Audit Commission report (Audit Commission, 2000) also concluded that in some areas unsuitable equipment was often provided, e.g. the provision of wheelchairs, and that users were often not asked basic questions about the support that they needed from such aids. The report recommended that users needed to be more closely consulted in service planning and delivery in order to ensure that services better match the need of their clients. Romich (1993) has also reported that there is often a lack of awareness of available products by users, carers and rehabilitation professionals, and there is little evidence to suggest the situation has changed significantly in Europe since then.

## 11.4 THE DEVELOPMENT OF THE USER*FIT* METHODOLOGY

The desire to create a user-centred design methodology for the European assistive technology sector gave rise to the original concept of the USER project. This was funded through the European Union's TIDE (Telematics for the Integration of Disabled and Elderly people) Programme as a support action to provide advice and resources for the other EC funded projects working in the advanced technology sector. The project drew largely from concepts and techniques developed in studies of human–computer interaction (HCI) and usability engineering, which were considered to have much to offer assistive technology, by providing methodologies and tools which facilitate a more user-oriented approach to design.

One of the outputs of this work was a design manual called USER*fit* (Poulson *et al.*, 1996), which aimed to encapsulate good practice and effective methods for design. USER*fit* offers a toolkit (see Figure 11.1) which provides guidance on user involvement and user-centred design, methods to achieve it and a range of helpful recommendations and design advice. Because USER*fit* is a modular toolkit, it is flexible and adaptable to different circumstances and can be used on its own or adapted to fit other design workframes. Unlike most other methods, however, USER*fit* covers not only the design of the product itself but also other, wider factors that can dramatically affect the success or failure of a product—for example, factors such as the environment in which the product will be used, training and support in the use of the product, and maintenance and after-sales service. These are important issues for any prospective purchaser of a product, but for the assistive technology market in particular they are critical.

At the heart of USER*fit* is the concept of usability—we say that a product is usable if it can perform effectively, efficiently, safely and comfortably the function for which it is being designed. If products and services do not have the necessary usability characteristics, safe and efficient operation may be seriously compromised. In consequence, it is very important for developers to take into account the characteristics of users, the things they do and want to do and where and when they will do them.

The USER*fit* Handbook consists of:

- A guide on the subject of user-centred design, usability, the principles of user involvement and the significance of user, activities and context characteristics for assistive technology
- The USER*fit* Methodology: a set of summary tools to collate, analyse, evaluate and develop information to build a specification, along with worked examples
- Descriptions of different specific design techniques for data capture and evaluation illustrating when and how to use them in the AT sector
- A collection of design prescriptions and recommendations concerning the design for assistive technology drawn from the scientific and technical literature
- A general information section containing a bibliography and information on useful sources of design information.

The USER*fit* Methodology is structured as shown in Table 11.1 (Poulson *et al.*, 1996).

A key aspect of this methodology is that it forces design issues to be made explicit. It makes designers, especially those who work in multidisciplinary teams, ask the right questions and justify and document any design assumptions or decisions they have made, either about the technology or about its users (Poulson

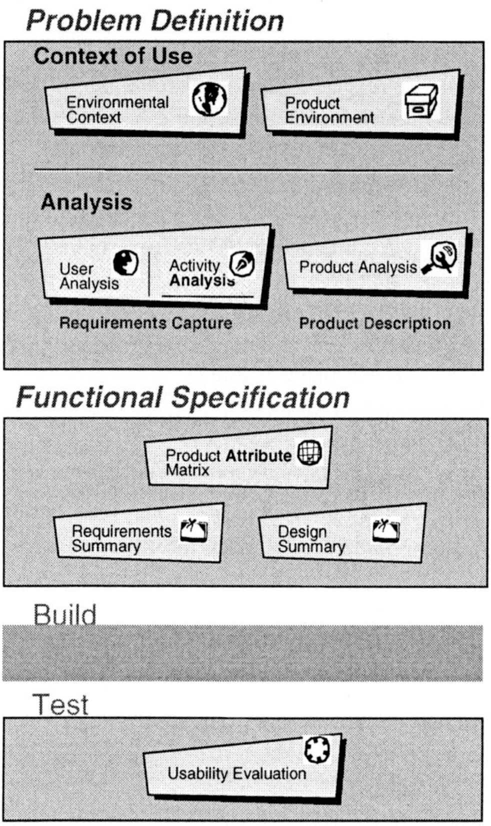

Problem Definition

Context of Use

Environmental Context

Product Environment

Analysis

User Analysis

Activity Analysis

Product Analysis

Requirements Capture

Product Description

Functional Specification

Product Attribute Matrix

Requirements Summary

Design Summary

Build

Test

Usability Evaluation

**Figure 11.1** Overview of the USER*fit* Methodology

and Richardson, 1998). However, the methodology is designed to be modular and can be visited where and when needed.

The tools making up USER*fit* are briefly described below, although for more detailed information about the elements, the reader is referred to the USER*fit* Handbook itself (Poulson *et al.*, 1996), or to a summary in Poulson and Richardson (1998).

## 11.5 APPLICATIONS OF THE USER*FIT* APPROACH

Since publication in 1996 approximately 2,400 copies of USER*fit* have been distributed. The majority of the material contained in the USER*fit* manual can also now be downloaded from a Web site supported by the EC-funded INCLUDE project. Access to the INCLUDE project site can be made from www.stakes.fi/include/

**Table 11.1** Elements of the USER*fit* methodology

| USER*fit* tool | Objectives |
|---|---|
| Environmental context | Provides a high-level summary of the product, covering such issues as the initial justification for it, who its users are likely to be, who will purchase it. |
| Product environment | Summarises what is known about the support environment for the product (including likely training, documentation, installation, maintenance and user support). |
| User analysis | Identifies the range of people who should be considered in the development of the product and describes in detail their attributes. |
| Activity analysis | Identifies and describes the range of activities that people will engage in when using the product and the implications that these will have for product design. |
| Product analysis | Summarises the functional aspects of the product as they are understood and lists these as operational features. |
| Product attribute matrix | Summarises the match between emerging functional specifications and product attributes inferred from user and activity analysis. |
| Requirements summary | Summarises the design features identified through user and activity analysis and their degree of match to user requirements. |
| Design summary | Summarises in more detail the functional specification for the product and its operational details. |
| Usability evaluation | Summarises plans for evaluation along with objectives, methods to be used and evaluation criteria. Also documents the degree of match between evaluation criteria and the results of evaluation activities. |

These materials can be obtained from the area of the site devoted to user-centred design methods and include all of the original USER*fit* manual apart from the section on design prescriptions (which could not be included for reasons of copyright).

The demand for USER*fit* has largely come from the assistive technology sector but there have also been requests for the handbook from institutes of higher education. For example, Abascal and Nicolle (2000) report the effective use of the methodology in teaching design students about usability issues and report that the methodology has been used with postgraduate students since 1997 in two Spanish universities. There is also some evidence that the handbook is beginning to be used in European research and development projects and a number of organisations are known to be exploring USER*fit*'s applicability to other application areas. For example, a survey was carried out at Loughborough University (Whitlock, 1998) involving a sample of 114 UK developers and academics based in the UK who had requested the USER*fit* manual. A total of 16 of the 25 who responded to the postal questionnaire had read some part of the USER*fit* manual, five had made use of the methodology and eight had made use of the tools and techniques. Over 90% of the sample were of the opinion that USER*fit* had been of some value to them. A more detailed analysis of these findings revealed that the component of the methodology covering user analysis was perceived to be the most valuable and that task analysis was rated as the most useful technique described. The inference drawn from this is that techniques to assist in user analysis and task definition are particularly needed in this sector and that in some part at least USER*fit* has been able to contribute to satisfying this demand. Research projects within Loughborough University have also added to an understanding of the manual's range of application and the manual is currently being used for the training of design and ergonomics undergraduates.

USER*fit* has also been deployed in the design of services in this sector (Goffee, 1996). This study evaluated the effectiveness of the USER*fit* methodology in helping to produce a set of requirements for the providers of a telemarketing service set up by the Enham Trust, a UK-based charity providing training and employment for people with disabilities. Even though USER*fit* was aimed at product development and not services, it raised a number of key issues and stimulated debate, for example with regard to software design and the wide range of training needs of the employees. This again emphasises the value of a structured framework which encourages questioning and communication. Foster and Solberg (1997) also report using a modified version of the USER*fit* methodology in the ARIADNE project, and other EC funded projects have also applied it in their work.

## 11.6 LIMITATIONS OF THE USER*FIT* APPROACH

USER*fit* is a framework for user-centred design and assists in the process of refining product specifications to take users into account. Particular emphasis is placed on an analysis of the design problem space, before commitment to a specific design solution is made. The potential danger of such an approach is that if care is not taken, it can lead to 'analysis paralysis' where increasingly large amounts of resources are spent in understanding the problem rather than solving it. Knowing when enough analysis has taken place is a matter of judgement, and design methodologies such as USER*fit* cannot easily provide advice on such issues. USER*fit* can assist in the process of design but is no substitute for the intuitions and experiences of good designers. USER*fit* also provides a framework to assist the designer by ensuring that topics considered important for usability are considered by design teams, though again judgement is needed on the part of designers in determining which of these to give priority in solving a particular design problem.

All design is about working within limitations and making compromises between conflicting requirements. Some of these requirements can also be described in terms of design guidelines which are summaries of accepted good practice. Guidelines are treated within USER*fit* in very much the same way as any other requirement and whilst USER*fit* helps in the process of collating such design requirements and objectives, it does not explicitly provide tools that can help in the resolution of conflicting demands. Again judgement by designers or design teams is required in order to resolve such conflicts.

At a purely operational and practical level, it also has to be acknowledged that feedback from users of USER*fit* indicates that some changes to USER*fit* are needed if its effectiveness as a design resource is to be improved. The full methodology is perceived as being somewhat difficult to follow, and the size and bulk of the manual is also a disincentive for use. One major criticism is that there are too many paper-based forms to complete when using the methodology and that a simplified methodology with an on-line tool for completing design forms could improve the usability of the product considerably.

## 11.7 CONCLUSIONS

General trends towards a greater degree of generic product development, a greater emphasis placed on the importance of direct user involvement and the value of using multidisciplinary design teams all have implications for assistive technology product design. New design methods and approaches are likely to be needed in order for design in this sector to accommodate these changes. The USER*fit* design methodology and handbook is one step towards achieving this objective.

The concepts and techniques developed in HCI and usability engineering have been demonstrated to be of value for assistive technology design. However, experience shows that this creates its own challenges, as those with disabilities differ considerably in their attributes and capabilities compared to a more able population. Such variability makes generalisation much more difficult to achieve.

It is difficult to assess fully the impact of the USER*fit* methodology and supporting materials, as the handbook has only been available in the past few years and to a relatively limited audience. The limited feedback obtained has demonstrated that many working in the assistive technology sector believe the manual to be a useful design resource but that reservations have also been voiced regarding practical aspects of use. Concern has been raised that the methodology as a whole is unwieldy to use and a more realistic scenario for use might be for developers to apply only those aspects which they found to be useful in a given design scenario. This vision is consistent with the philosophy behind the development of USER*fit*, as it was designed also to be applicable in a modular fashion.

However, we needed to start somewhere and we feel that the present USER*fit* Handbook is a step in the right direction towards promoting the concept of user-centred design and user involvement in the assistive technology sector. Now we are seeking opportunities to improve the Handbook so that it will not only be more useful to designers in promoting *Design for All* but will also be more usable in itself.

CHAPTER TWELVE

# The ISO Approach to the Development of Ergonomics Standards for Accessibility

Jan Gulliksen, Susan Harker and John Steger

## 12.1 INTRODUCTION

With the increasing interest in designing usable systems and the increasing recognition of the need to provide accessibility for people with specific potential requirements, the member bodies of ISO Technical Committee 159, Ergonomics, have adopted a work item on software accessibility which has resulted in the development of a Technical Specification, ISO 16071 *Ergonomics of human–system interaction—guidance on software accessibility* (International Organisation for Standardisation, 2000). This work represents a response to the need to provide as wide a range of potential users as possible with access to software products and applications. Recognition of the significance of ensuring that people with disabilities are able to play a full role in the job market is reflected in the increasing number of legal requirements (such as the Americans with Disabilities Act 1992, new legislation requiring US Government procurement to ensure accessibility, etc.).

It is important to address the growing population of users of IT systems who need extra support to be able to handle and find the technology useful. People are different and therefore these differences need different types of support to give the users an equal opportunity to access electronically available information and public services as well as to be able to participate in working life. The *Nordic Guidelines for Computer Accessibility* recognise four different categories of users who potentially would benefit from these types of requirements (Nordic Cooperation on Disability, 1998):

- People with physical, sensory or cognitive impairments, either present at birth or acquired by illness, accident or age
- The growing number of elderly people (25% of the population in Europe will be aged 60 or above in the year 2020)
- Temporary disabilities, e.g. broken arm, pregnancy, forgotten glasses at home, etc.
- Disabling environments, which is something that in itself should not be recommended, but in situations where it cannot be avoided, specific support is needed.

## 12.2 INTERNATIONAL ORGANISATION FOR STANDARDISATION (ISO)

The International Organisation for Standardisation (ISO) is a worldwide federation of national standards bodies from some 130 countries, one from each country. It is a non-governmental organisation established in 1947. The head office is in Geneva. The mission of ISO is to promote the development of standardisation and related activities in the world with a view to facilitating the international exchange of goods and services, and to developing co-operation in the spheres of intellectual, scientific, technological and economic activity. ISO's work results in international agreements, which are published as International Standards.

ISO standards (www.iso.ch/infoe/intro.htm) are developed according to the following principles:

- Consensus—The views of all interests are taken into account: manufacturers, vendors and users, consumer groups, testing laboratories, governments, engineering professions and research organisations
- Industry-wide—Global solutions to satisfy industries and customers world-wide
- Voluntary—International standardisation is market-driven and therefore based on voluntary involvement of all interests in the marketplace.

The International Standard described in this document has been developed in the form of a Technical Specification by ISO TC 159/SC 4/WG 5 *Software ergonomics and human computer dialogues*. Working Group 5 is the group responsible for the development of the software parts of ISO 9241 *Ergonomic requirements for office work with visual display terminals (VDTs),* that is parts 10 through 17. WG 5 is also currently preparing ISO 14915 *Multimedia User Interfaces*.

The initiative for the new work item was provided by the technical group responsible for covering software ergonomics in the United States, that is, the American National Standards Institute/Human Factors and Ergonomics Society (ANSI/HFES) Committee on Human Computer Interaction Standards. This group has been working actively in the area of accessibility for a number of years and has recently completed the ballot process in the United States for a document which is the second part of a multi-part standard on human factors engineering of software interfaces. This accessibility document provides design recommendations for features and functions of computer operating systems, drivers, application services and applications that increase the accessibility of applications for users with disabilities. The ISO working group has been able to move forward with relative speed due to the valuable work undertaken within the ANSI/HFES Committee. Much of the guidance developed by this group has been included in the ISO document. However, the scope has been extended to reflect the consensus within the international group of experts responsible for the ISO work.

The decision to develop ISO 16071 *Ergonomics of human–system interaction—guidance on software accessibility* as a Technical Specification (TS), rather than proceeding directly to the development of an International Standard, was prompted by a wish to use the opportunity to include a wide range of material which has not traditionally been the subject of standardisation activities. A TS is a new format for ISO documents presented first in 1998. It is a normative document representing the technical consensus within an ISO committee. While being subject to the same process of voting and comments by the national bodies who are members of the committee, it offers the potential to include material which is relatively new and which it might be thought premature to include in an International Standard (IS). Following publication of a TS, provision is made to review the document with a view to converting it into an IS. The first review takes place within three years and, if the TS is not regarded as appropriate for approval as an IS within six years, it is withdrawn. It is the intention of the working group that response to the TS should be monitored following its publication and that feedback on its application should be used to prepare a revision of the text which can be put forward as an IS as soon as possible.

## 12.3 THE INTRODUCTION TO ISO TS 16071

The purpose of ISO Technical Specification 16071 is to provide guidance to developers on designing human–computer interfaces which can be used with as high level of accessibility as possible. Designing human–computer interactions to increase accessibility promotes increased effectiveness, efficiency and satisfaction for people who have a wide variety of capabilities and preferences. Accessibility is therefore strongly related to the concept of usability (ISO 9241 Part 11—*Guidance on Usability*).

The most important methodological approaches to increase the accessibility of a given human–computer interface are:

- Task-oriented design of user interfaces
- Customisation
- The use of human–centred design principles (ISO 13407—*Human centred design process for interactive systems*)
- Individualised user instruction and training.

The focus of ISO TS 16071 is the development of human–computer interactions to systems and products that will enable their use by the widest range of people with special needs. An important part of a human-centred design process for accessible systems is to develop human–computer interfaces to meet accessibility goals that can be quantitatively evaluated for a specific user or user category in a specified context of use.

ISO TS 16071 is based mainly on the prevalent knowledge of individuals with sensory and/or motor impairments in a work context. However, accessibility is an attribute that affects a large variety of capabilities and preferences of human beings in a range of different settings. These different capabilities may be the result of age, disease and/or disabilities. Therefore, accessibility addresses a widely defined group of users including:

- people with physical, sensory and cognitive impairments present at birth or acquired during life

- elderly people who can benefit from new IT products and services but who experience reduced physical, sensory and cognitive abilities
- people with temporary disabilities, such as a person with a broken arm or someone who has forgotten his or her glasses
- people who are experiencing difficulties in certain situations, such as a person who works in a noisy environment or has both hands occupied by other work.

Having a disability should be regarded as a natural element of the life cycle. Everyone can expect, during some period of life, to be affected by circumstances that make the access to and use of IT systems products and services potentially more difficult.

ISO TS 16071 recognises that some users will always need assistive devices to use a system. Therefore, it includes the capability of a system to connect and interact successfully with assistive technologies in the concept of accessibility. However, it does not address the specific issues associated with the design of interfaces to individual forms of assistive technologies.

ISO TS 16071 provides guidance for system design, appearance and behaviour. The goal of the guidance is to allow software to be used by as broad an audience as possible. In addition, guidance is provided on designing software to integrate as effectively as possible with common assistive technologies (e.g., speech synthesisers, Braille input and output devices) when they are available. Incorporating accessibility features early in the design process is relatively inexpensive compared to the cost of modifying products to become accessible.

ISO TS 16071 addresses the increasing need to consider the social and legislative demands to ensure accessibility by removing barriers that prevent people from participating in life activities including the use of environment, services, products and information. Designing software user interfaces for accessibility increases the number of people who can use computer systems by taking into account the varying physical and sensory capabilities of user populations. Designing for accessibility benefits disabled users by making software easier for them to use or by making the difference between whether or not they are able to use the software at all.

Many accessibility features also benefit users who are not currently experiencing disability by enhancing usability and providing additional customisation possibilities. They may also help to overcome temporary deficits (e.g., a broken arm or hand). ISO TS 16071 benefits designers and suppliers by expanding the number of potential users (and thus sales for their product) and often by making the product compliant with legal requirements for accessibility. It benefits companies buying software by expanding the number of employees who may use the software.

Accessibility may be provided by a combination of both hardware and software. Assistive technologies typically provide specialised input and output capabilities not provided by the system. Software examples include on-screen keyboards that replace physical keyboards, screen magnification software that allows users to view their screen at various levels of magnification, and screen reading software that allows blind users to navigate through applications, determine the state of controls and read text via text-to-speech conversion. Hardware examples include head-mounted pointers that replace mice and Braille output devices that replace a video display. There are many other examples not listed here. When users provide add-on assistive software and/or hardware, usability is enhanced to the extent that systems and applications integrate with those technologies. For this reason, operating systems may have to provide 'hooks' or other features to allow software to operate effectively with add-on assistive software and hardware as recommended in these guidelines. If systems do not provide support for assistive technologies, the probability increases that users will encounter problems with compatibility, performance and usability. At the same time, if software applications do not use system-provided mechanisms (such as customisation for colour, font and audio, or system routines for keyboard navigation and text input), then users can find their access blocked.

## 12.4 SCOPE OF ISO 16071

The scope of an ISO Standard or a Technical Specification is particularly important in providing a clear statement of the issues addressed both in terms of the form in which the guidance is offered and topics which are covered. The Scope of TS 16071 is set out below: (ISO 16071)

> *"ISO Technical Specification 16071 provides guidance on design of accessible (work, home, education) software. It covers issues associated with designing accessible software for the widest range of visual, hearing, motor and cognitive abilities, including people who are elderly and temporarily disabled. This Technical Specification addresses software considerations for accessibility that complement general usability design covered by ISO 9241 and ISO 13407.*
>
> *The Technical Specification covers accessibility of computer operating systems and applications. While it does not address all accessibility issues of platforms and application domains such as Web pages, multimedia, personal digital assistants (PDA) and kiosks, many of the recommendations are applicable in these areas. It does not apply to software used primarily for entertainment purposes (i.e., games). It does not provide recommendations for the design of hardware.*
>
> *The Technical Specification is aimed at reducing the need for add-on assistive hardware and software technologies while promoting increased usability of systems in combination with assistive technologies, when they are required. It does not cover the behaviour or requirements for assistive technologies themselves (including assistive software)."*

## 12.5 DEFINING ACCESSIBILITY IN RELATION TO USABILITY

Experience in the development and use of ISO 9241 Part 11—*Guidance on Usability* demonstrates the benefits of having a clear and succinct definition of the concept being dealt with. ISO 9241:11 defines *usability* as:

*The extent to which a product can be used by specified users, to achieve specified goals, with effectiveness, efficiency and satisfaction, in a specified context of use.*

The advantage of such a definition is both that it defines usability to be a quantitatively measurable concept and that it emphasises the non-functional demands of usability that are so important for the final interpretation of the concept.

Experience with the definition of usability has led to the conclusion that it is desirable to define accessibility in relation to usability as a measurable entity. Therefore ISO TS 16071 defines *accessibility* as:

*The usability of a product, service, environment or facility by people with the widest range of capabilities.*

This definition implicitly indicates that it is measurable through its relation to usability. However, it is extended to cover more than just products (as in the definition of *usability*), including service, environment or facility as well. Also it is apparent that the level of accessibility will be dependent upon having identified 'specified users' who encompass the widest possible range. This definition can be compared to the definition from ANSI/HFES 200 *Software User Interfaces–Accessibility*, which addresses the concept in terms of the attributes of the artefact being delivered rather than the measured extent to which it can be used:

*The set of properties that allows a product, service or facility to be used by people with a wide range of capabilities, either directly or in conjunction with assistive technologies. Although 'accessibility' typically addresses users who have a disability, the concept is not limited to disability issues.*

This definition suggests that accessibility has a binary nature, i.e., something is accessible or not. This does not correspond with the underlying character of the specific guidelines provided in both the ANSI/HFES document and in the ISO Technical Specification. It is, therefore, more appropriate to sharpen the definition to be quantitatively measurable and comparable to the definition in ISO 9241 Part 11.

These approaches to the definition of accessibility can be compared with the definition used by the World Wide Web Consortium in the *WAI–Web Content Accessibility Guidelines* 1.0 (W3C, 1999):

*Content is accessible when it may be used by someone with a disability.*

Such a definition would attribute accessibility to content that can be accessed by any person with a disability, without obliging any individual or organisation claiming to have produced accessible content to identify the user characteristics of those groups of people with disabilities who should be able to access it. Measurable accessibility should be measured in terms of users, task, context and products. To be able to compare the accessibility one must only vary one of these terms at a time, e.g.:

$Accessibility_1 = f(user_1, task_1, context_1, product_1)$
$Accessibility_2 = f(user_1, task_1, context_1, product_2)$

A consequence of this is that to be able to measure and compare accessibility between two different systems, we must only vary one aspect at a time. This means new methodological challenges and difficulties since it is very difficult in practice to compare the accessibility features in different contexts of use. The users' personal characteristics may also change between, for example, two studies of the same individuals, performing the same tasks.

## 12.6 RATIONALE AND BENEFITS OF ISO 16071

Accessibility is an important consideration in the design of products, systems, environments and facilities because it affects usability by people with the widest possible range of capabilities.

Accessibility can be improved by incorporating features and attributes known to benefit the users with specific special requirements. In order to determine the achieved level of accessibility, it is necessary to measure the performance and satisfaction of users working with a product or interacting with an environment. Measurements of accessibility are particularly important in view of the complexity of the interactions with the user, the goals, the task characteristics and other elements of the context of use. A product, system, environment or facility can have significantly different levels of accessibility when used in different contexts.

Planning for accessibility as an integral part of the design and development process involves the systematic identification of requirements for accessibility, including accessibility measurements and verification criteria within the context of use. These provide design targets that can be the basis for verification of the resulting design.

The approach adopted in ISO Technical Specification 16071 has benefits, which include:

- The framework can be used to identify the aspects of accessibility and the components of the context of use to be taken into account when specifying, designing or evaluating the accessibility of a product
- The performance (effectiveness and efficiency) and satisfaction of the users can be used to measure the extent to which a product, system, environment or facility is accessible in a specific context
- Measures of the performance and satisfaction of the users can provide a basis for the comparison of the relative accessibility of products with different technical characteristics, which are used in the same context
- The accessibility planned for a product can be defined, documented and verified (e.g., as part of a quality plan).

## 12.7 TARGET AUDIENCE

Who will be using such an ISO document? Today, many organisations have a user-centred design group within the organisation to take responsibility for accessibility issues.

ISO Technical Specification 16071 serves the following types of users:

- the *user–interface designer*, who will apply the guidance during the development process
- the *developer*, who will apply the guidance during design and implementation of system functionality
- the *buyer*, who will reference the Technical Specification during product procurement
- *evaluators*, who are responsible for insuring products meet the recommendations of the Technical Specification
- *designers of user–interface development tools and style guides* to be used by interface designers.

The ultimate beneficiary of this Technical Specification will be the end-user of the software. Although it is unlikely that the end-users will read the Technical Specification, its application by designers, developers, buyers and evaluators should provide user interfaces that are more accessible. The guidelines concern the development of software for user interfaces; however, those involved in designing the hardware aspects of user interfaces may also find them useful.

It is undoubtedly the case that International Standards offer advantages in establishing an increased awareness and impact in relation to the problems they address. Hence the development of the Technical Specification and ultimately the IS should provide further weight to the case for ensuring that software products and systems are designed to make them accessible by the widest possible range of users. Having an International Standard as compared with just having access to guidelines published by special interest groups is advantageous. Standards, because of the process they have been through, are generally regarded as carrying more weight and they do give specifiers and suppliers a reference point to use when negotiating about the properties required in the product or system.

## 12.8 WHAT TYPES OF CAPABILITIES DO WE WANT TO ADDRESS?

The user characteristics of people with any given disability vary significantly just as any other heterogeneous population. The descriptions contained here provide only an outline of the issues typically encountered by individuals with various disabilities and do not constitute a comprehensive list. People may concurrently experience more than one of the disabilities outlined below. The needs of people who have such combinations of disabilities are covered in several cases by the overlap across guidelines.

The following capabilities are specifically addressed in this standard:

- users who are blind

- users who have low vision
- users who are deaf
- users who have hearing impairments
- users who have physical impairments
- users who have cognitive impairments
- users who are elderly
- users who have temporary disabilities
- users who have multiple disabilities
- users who have environmental disabilities.

### 12.8.1 Issues commonly encountered by users who are blind

The primary issue for users who are blind is how to obtain information provided by visual presentation, how to navigate among objects on screens, how to identify those objects, and how to control focus, navigation and other functions via the keyboard.

Many people who are blind from birth learn Braille, and many who become blind later in life often rely on additional auditory methods to obtain information.

Many users who are blind often have some vision and primarily interact with computers through 'screen readers'—assistive software that can provide spoken or Braille information for windows, controls, menus, images, text and other information typically displayed visually on a screen.

Considerations for these users follow from the characteristics of interactions mediated by screen readers. To the extent that interactions depend on understanding a spatial metaphor for navigation or seeing graphically represented objects, users who are blind are more likely to encounter difficulties and normally keyboard navigation is an essential function.

In addition, because many users who are blind are reading screens by means of synthesised speech output, they may find it difficult or impossible to attend to auditory outputs that occur while they are reading.

### 12.8.2 Issues commonly encountered by users who have low vision

The issues commonly faced by users who have low vision include loss of visual acuity, colour perception deficits, impaired contrast sensitivity and loss of depth perception.

People who have low-vision use varying means of increasing the size, contrast and overall visibility of visual displays depending upon their visual needs. Common assistive technologies include use of oversized monitors, large fonts, high contrast and hardware or software magnification to enlarge portions of the display.

When interacting with computers, these users may not detect size coding, may have difficulty with font discrimination and may encounter difficulties locating or tracking interface objects such as pointers, cursors, drop targets, hot spots and direct manipulation handles.

Additionally, both blind users and users who have low vision experience difficulties when required to read very small displays, such as those on printers, copiers, ticket machines and automatic teller machines (ATMs).

### 12.8.3 Issues commonly encountered by users who are deaf or hard of hearing

Issues for users who are deaf or hard of hearing who retain some functional hearing include the inability to discriminate frequency changes, decreased frequency range and dropout, difficulties localising sounds, and difficulty picking up sounds against background noise.

Users who are deaf or hard of hearing may or may not use electronic hearing aids, depending on the nature and extent of the hearing impairment. If it is available in the operating system, they may use the 'ShowSounds' feature that notifies software to present audio information in visual form.

When interacting with computers, these users may have trouble hearing sounds of certain frequencies, or of low volume. Sound customisation is key to providing them with access.

### 12.8.4 Issues commonly encountered by users who are deaf

In addition to a general inability to detect auditory information, the issues commonly faced by users who are deaf may include the inability to produce speech recognisable by voice input systems and experience with a national language only as a second language (sign language often being the primary language for people who are born deaf or who become deaf at an early age).

If it is available, users who are deaf will typically use the 'ShowSounds' feature that notifies software to present audio information in visual form.

When interacting with computers, these users will encounter problems if important information is presented only in audio form. Many of these issues apply to any user in contexts where sound is masked by background noise (e.g., machine shop floor) or where sound is turned off or cannot be used (e.g., a library).

### 12.8.5 Issues commonly encountered by users who have physical impairments

The issues commonly faced by users who have physical impairments often follow from physical limitations, including poor co-ordination, weakness, difficulty reaching and inability to move a limb.

Users with physical impairments may or may not use assistive technologies and the variety of hardware and software they employ is too large to describe in detail in this space. A few examples, however, include eye-tracking devices, on-screen keyboards, speech recognition and alternative pointing devices.

Some users may have difficulty directly manipulating objects, using modifier keys, using pointing devices and performing actions that require precise movement or timing. Other users may have tremors that cause difficulty in moving to a target. The extreme variation in needs and capabilities among this user population means that customisation of input parameters and timing is extremely important for effective access.

### 12.8.6 Issues commonly encountered by users who have cognitive impairments

The issues commonly encountered by users who have cognitive disabilities involve difficulties receiving information, processing it and communicating what they know. People with these impairments may have trouble learning new things, making generalisations and associations and expressing themselves through spoken or written language. Attention deficit hyperactivity disorders make it difficult for a person to sit calmly and give full attention to a task.

The issues commonly faced by users who have dyslexia are difficulties reading text that is presented in written form and difficulties in producing written text.

Reading difficulties are best supported by having text that is highlighted and read out loud or by providing 'easy-reading' versions of the texts. Users without reading difficulties also benefit from easy-reading versions of written text.

Providing synthetic speech for what is to be written best supports writing difficulties.

### 12.8.7 Issues commonly encountered by users who are elderly

Elderly users are progressively limited in their ability to use and access human–computer interfaces due to the multiplicative effects of combinations of visual, hearing, cognitive and motor impairments that to varying degrees might come with increasing age.

Sometimes the awareness of diminishing capabilities is a concern for elderly users. Therefore, built-in accessibility of products contributes to removing the stigma of special aids or modifications. Elderly persons do not want to have their age regarded as a disability.

### 12.8.8 Issues commonly encountered by users who have temporary disabilities

Temporary disabilities are often of a physical nature (e.g., broken arm). These users seldom adopt efficient skills in learning to cope with their disability. It is, therefore, important to make the accessibility features for these disabilities easy to find and learn to master.

Temporary disabilities might also be caused by repetitive strain caused by poor ergonomics and intensive use of the computer system. It is then important that this injury can be relieved through support that can be provided to the user via the system. For example, improving the design of the laptop computer so that it can be opened with one hand also increases the usability for all users.

### 12.8.9 Issues commonly encountered by users who have multiple disabilities

There are not just a few categories of disabilities—rather the range of different accessibility needs varies just as much as the combinations of certain degrees of disabilities. For example, an individual with a cognitive impairment might also have low vision.

Several of the guidelines for addressing a specific disability might be contradictory. For example, auditory output of written text is not a support for the deaf–blind. It is, therefore, important that the support for these forms of multiple disabilities is individualised for the specific user and task.

### 12.8.10 Issues commonly encountered by users who have environmental disabilities

Environmental disabilities occur when specific features of the work environment cause difficulties in perceiving signals from the computer. Such situations include difficulty hearing signals from the computer when working in a noisy environment. These situations must be regarded as disabling the user to fulfil the task with the aid of the computer. Although this may not be a software requirement, the immediate solution should be to improve the environment, and in situations where this cannot be done such as in airports, to provide redundant presentations of essential information—i.e., by providing several modes of feedback to the user in case the environment prevents the user from perceiving one particular mode.

## 12.9 WEB ACCESSIBILITY INITIATIVE (WAI) GUIDELINES

The W3C has provided one of the most widely used documents on accessibility. Its coverage is of course directed at interfaces and interactions based on use of the World Wide Web, and the scope of TS 16071 was formulated in the knowledge that this guidance was available. However, the ISO working group did have an extended discussion concerning the extent to which the material in the WAI guidelines could be regarded as stable and how a standard could reference guidelines that only are provided on line. The conclusion of the debate was that it would be appropriate to ensure that any guidance on accessibility that is generally applicable should appear in the TS. The guidelines provided in W3C, 1999, were reviewed and grouped into the following categories:

- guidelines that are already covered by ISO 9241
- guidelines that are not of a general nature
- guidelines that should be included in the document.

We found that most of the guidelines provided in the WAI document were standard ergonomic guidelines already covered in ISO 9241 and others. ISO TS 16071, used together with other software accessibility standards, covers all of the WAI guidelines on software accessibility.

## 12.10 DISCUSSION

ISO TS 16071 *Ergonomics of human–system interaction—guidance on software accessibility* will be sent for publication by the end of 2000. The working group will then initiate a discussion about how to move the TS forward for approval as an International Standard. An important part of this review will involve assessing the reactions of the intended users of the document to its current coverage and content. It is recognised that there are areas where further design guidance is needed, but if current research and application have not yet reached a position where there is a stable knowledge base, it is unlikely to be appropriate to include this material in a standard. Indeed, one question which may arise is the extent to which the guidance in the current document is sufficiently robust to move from the status of a TS to an IS. This could result in some content being removed in the final version of the IS. However, it is also appropriate to consider whether there is scope for emerging material to be included in another TS running in parallel with the IS. There are also the questions about the use of other forms of ISO output, such as the Publicly Available Specification, to provide supporting tools such as checklists to be used for evaluation purposes.

In the further development of an International Standard based on the Technical Specification, we believe that several issues that are new to the ISO way of performing standardisation need to be addressed, for example, the following:

- *Strengths and weaknesses of standards.* It is important to strive to make the presentation of standards more attractive. The formal, international process of adopting standards brings weight, but it also forces them to conform to a rather unusable format and stilted form of words to avoid the possibility of ambiguity that may arise with a more flowing style. Large organisations who use standards will probably have the time and resources to interpret them for use in their particular design environment. This is obviously a much greater problem in small organisations, which is why training has an important part to play. This is a somewhat neglected area in relation to the dissemination of standards.
- *Rigour versus speed.* The formal voting procedures of ISO are, of course, a limitation to the prospect of having dynamically updated documents, such as the WAI guidelines. However, for that very same reason, these procedures serve their purpose as they present knowledge that has been through a very tough and sound review process.
- *Checklists and examples.* We know from practical experience that the only parts of a standard that are used directly in the design process are checklists, which are used to tick off whether an artefact or process fulfils the guidelines it exemplifies, and examples, which can either be copied or serve as inspiration to the user of the standard. A common phase in the evolution of

standards is for the working group to provide more examples in response to comments and requests made at the time of voting. Subsequently, these examples tend to be removed as the voting procedure proceeds because it is always difficult to get consensus on the usefulness and validity of examples. One advantage is that those that survive this process have greater rigour. Different checklists addressing the same problems could have different form and content but achieve the same goals. In standards one could expect that it would be difficult to get any single checklist agreed as part of the main provisions of a standard. In ISO 9241 Parts 12–17 the checklist is specifically stated to be an example. According to the ISO procedures the Publicly Available Specification would offer an appropriate medium for publishing alternative checklists.

- *One-stop shopping.* As a software developer concerned with ergonomics and human–centred design, one currently needs to have knowledge on far too many different standards, even considering only those produced by the ISO Ergonomics Technical Committee. The software parts of ISO 9241 and the multimedia guidelines from the four-part standard ISO 14915, as well as process standards from ISO 13407, are the most commonly used and are still only a fragment of the existing requirements and guidelines in ISO and IEC documents relevant for software developers. ISO TC 159/SC 4 is currently revising its documents and the structure of ISO 9241 and related standards with the intention of specifically addressing these issues. In particular, the aim is to group the standards in a way which will address the needs of the specific users. What outcome this process will have is uncertain at this moment.

Perhaps even more important than the Standard in itself is the process of creating the Standard. Working drafts, which generally contain much more information than the final version of the Standard will have, are widespread—and even though they should not be referenced, they often are. Thus the process of development may be as influential in changing attitudes and practice as is the direct application of the content. This is one of the reasons why the Working Group is pursuing the development of this Technical Specification into a Standard. Even though it is felt that there is much more knowledge within accessibility to be gained, the increasing need for these types of documents to be visible in the market makes it necessary to force the pace of development so that they reach the public domain as soon as possible.

## ACKNOWLEDGEMENTS

The authors wish to acknowledge the entire subgroup on accessibility from ISO Technical Committee 159/Sub-committee 4/Working Group 5. The input provided by the Human Factors and Ergonomics Society in terms of their outline of the accessibility part of ANSI 200 is a significant contribution to the making of the Technical Specification.

# Part 5

# Guidelines for Specific Application Areas

CHAPTER THIRTEEN

# Guidelines for Telecommunications

Jan Ekberg and Patrick Roe

## 13.1 INTRODUCTION

Older and disabled people form a very significant proportion of the telecommunications market—a proportion with increasing disposable income. By the year 2020, a quarter of the population in Europe will be over the age of 60 and many of them will face some degree of difficulty in using telecommunications equipment if this is not properly designed. The challenge for manufacturers and service providers is to make their products accessible to this significant sector of the market.

This chapter highlights some aspects of good design of various telecommunications devices and services and gives references to more detailed guidelines. By applying such good design we can fulfil the requirement, voiced by the Secretary General of the International Telecommunication Union in his message to Telecom 99, namely that "It is essential that global telecommunications development should be equitable."

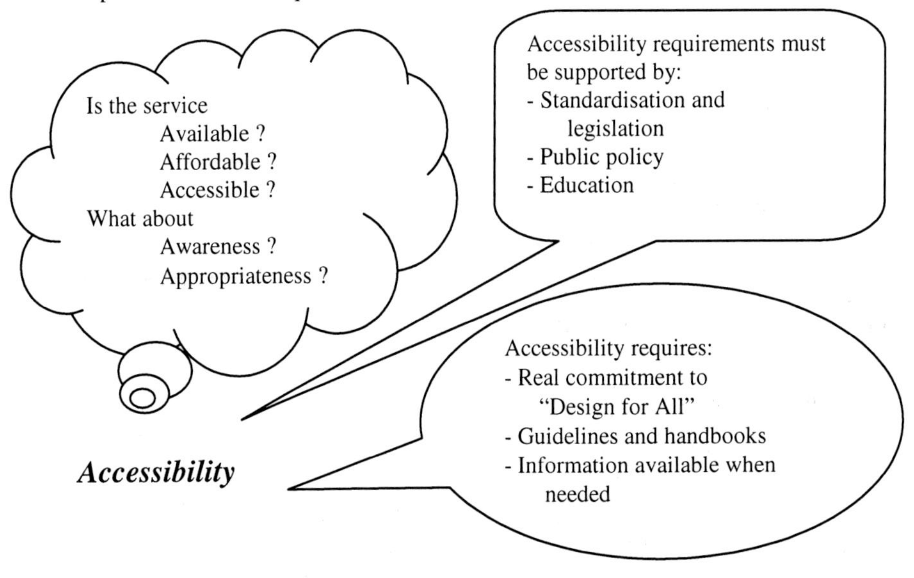


**Figure 13.1** Accessibility is not just a matter of guidelines

## 13.2 PREREQUISITES FOR UPTAKE OF NEW TECHNOLOGY

Many human beings are not able to profit from the new services and devices being developed today. The ISPO/Promise project has defined the five areas that influence the uptake of a new device or service in the following way (see Figure 13.1 and www.stakes.fi/promise/book/pr00intr.htm#5):

**Availability**
Wide availability of equipment and on-line services is a fundamental prerequisite. On the one hand, users must have the necessary terminal equipment (e.g., PCs and modems, text telephones or videophones), telecommunications connections (e.g., ordinary telephone lines, ISDN lines or interactive Cable TV services), and subscriptions to services such as the Internet. On the other hand, service providers must offer the necessary on-line services (e.g., teleshopping, electronic books or distance learning). Information society initiatives must therefore define ways of ensuring that individuals have access to the equipment and connections that they need, and that service providers are encouraged to provide useful and desirable services.

**Accessibility**
Given the various disabilities that the different user groupings may have, the extent to which products and services are designed to be accessible to all is another key factor. Web pages must be designed to be accessible to people with visual impairments, for example, and the audio content of TV and other audio-visual services must be captioned and/or signed to be accessible to people with hearing impairments. Without full accessibility, significant numbers of potential users will be excluded. Information society initiatives must therefore give the highest priority to the promotion and implementation of 'Design for All' or 'Inclusive Design' to ensure that everyone can have access.

**Affordability**
If older people and disabled people are to benefit from the opportunities presented by the information society, the products and services must not only be available and accessible but they must also be affordable. Disabled people often have low incomes and many older people are still at risk of poverty—therefore public financial support is an important factor. Information society initiatives must therefore actively address the financial dimension and ensure that lack of income does not exclude the participation of those who could benefit the most.

**Awareness**
One major barrier to an inclusive information society is the lack of awareness in public policy, in industry and in other sectors such as education, about the needs of disabled people and older people, and of the ways in which these needs can be catered for. From the point of view of disabled and older people themselves, of course, availability, affordability and accessibility of applications are not enough on their own to enable diffusion and take-up—potential users must also be aware of what is possible and be interested in taking up the opportunities. Therefore, the information levels and attitudes of older people and disabled people must be addressed in any promotional initiatives, as well as the levels of awareness and

willingness to innovate on the part of industry, social and health services, employers, educational institutions and other relevant sectors.

**Appropriateness**
Finally, it is important to assess the appropriateness of the various applications in particular circumstances. One of the main ways to do this is through user-involvement and it is imperative that older people and disabled people themselves are actively involved in the design, development and evaluation of new applications and services. One issue, in particular, that needs to be addressed is the fact that the information society can represent something of a double-edged sword for older people and disabled people in many cases. It can open up new links and new opportunities for contact and participation but, paradoxically, it can also result in new forms of social isolation if used inappropriately. Information society initiatives must therefore include social assessment as a central dimension.

In this chapter only the accessibility to telecommunications will be analysed in more detail, since this is where HCI guidelines are particularly relevant.

## 13.3 ACCESSIBILITY GUIDELINES ARE NOT ONLY FOR DESIGNERS

Providing researchers with information on accessibility and good design is not enough. The requirement that public telecommunications services and terminals have to be accessible has to be considered throughout the whole chain consisting of, for instance:

- research, development and manufacturing
- standardisation
- legislation
- public policy
- education.

Social, health and ethical aspects (like autonomy, integrity, self-esteem, justice, utility for all) must also be taken into account, for instance when installing new technology in the homes of elderly people. As always, it is not technology as such that creates problems, but the way it has been designed and the way it is used.

The European research actions COST 219 and COST 219bis have studied the accessibility of telecommunications services and devices and provided input to all parts of the chain mentioned above.

The activities started early in 1985 by the COST[1] Technical Committee Telecommunication (TCT), based on a report *Report and Proposal on Possible COST Activities Relating to the Provision of Telecommunications and Tele-informatics Facilities for Disabled People* from Mr Joseph Dwyer. This report underlined the need to continue the work started by the Nordic TELEMEDEL[2]

[1] COST stands for European Co-operation in the field of Scientific and Technical Research.

[2] TELEMEDEL collated information about how telecommunications could be used to alleviate problems encountered by disabled people.

initiatives in 1979 and the conference on text communication for the deaf (The Hague, 25–26 October, 1984).

The activities started at a time when usability issues and human factors had become a hot topic for industry. New technology, such as processors, made it possible to design user interfaces based on the priorities and preferences of the users. The telephone dial, for instance, that was earlier used because it could mechanically produce the pulses needed to control the selector in the telephone exchange, was replaced with the more user friendly key pad because the pulses needed could now be produced electronically after a single key press.

## 13.4 COST 219bis 'TELECOMMUNICATIONS: ACCESS FOR DISABLED ELDERLY PEOPLE'

COST 219bis is a European research project (www.stakes.fi/cost219) with official participants from Australia, Austria, Belgium, Denmark, Finland, France, Germany, Greece, Hungary, Ireland, Italy, The Netherlands, Norway, Portugal, Slovenia, Spain, Sweden, Switzerland, the United Kingdom and the United States.

The main objectives of the Action are:

- to ensure that the design of telecommunication services and equipment renders them accessible to elderly people and people with disabilities
- where it is not possible to ensure that the design of telecommunication services and equipment are accessible, to ensure that they are capable of being adapted
- to establish special services and equipment in case the above aims cannot be achieved in a successful way.

The work for improving accessibility in telematics services and devices in COST 219bis is based on the 'Design for All' or 'Universal Design' concept, also often referred to as 'Inclusive Design'. This part of the work includes:

- identification of the impact of technical, social and economic developments on elderly people and people with disabilities
- production of guidelines to meet the needs of the target groups and influence the development of services and equipment
- exchange and dissemination of results from research and specialists' experiences
- promotion of end-user involvement as well as involvement of industry, service providers, and policy and decision makers in the process.

The work has resulted in reports, for instance, on user requirements and alleviation to problems (Frederiksen *et al.*, 1991), on technical opportunities and implementations (Roe, 1995), and on legislation and standardisation (Stephanidis, 1994; Gjøderum, 1995). The glossy booklet (Gill and Shipley, 1997) on *Disaster or Opportunity? The Impact of Telecommunications Deregulation on People with Disabilities* could be mentioned as one of the publications aiming at making public policy aware of the problems in the area. COST 219bis has also actively supported the universal services issues (Lindström, 1998).

The 'Design for All' (www.stakes.fi/include/incd420.html) or 'Universal Design' principle is a philosophy or strategy based on the principle that products should be usable by as wide a range of the population as possible. Universal design is based on the notion that by ensuring that the least able can use a product, one maximises the number of potential users and also creates products and services which are easier for the more able to use. It can be argued that it is impossible truly to design for all, but the principle aims at eliminating the fewest possible customers. The Center for Universal Design has presented seven universal design principles (www.stakes.fi/include/pam2.html) to guide a wide range of design disciplines:

- Equitable use: the design is useful and marketable to any group of users
- Flexibility in use: the design accommodates a wide range of individual preferences and abilities
- Simple and intuitive use: use of the design is easy to understand, regardless of the user's experience, knowledge, language skills or current concentration level
- Perceivable information: the design communicates necessary information effectively to the user, regardless of ambient conditions or the user's sensory abilities
- Tolerance for error: the design minimises hazards and the adverse consequences of accidental or unintended actions
- Low physical effort: the design can be used efficiently and comfortably and with a minimum of fatigue
- Size and space for approach and use: appropriate size and space is provided for approach, reach, manipulation and use regardless of the user's body size, posture or mobility.

The 'Design for All' information in the COST219 listing of general requirements (www.stakes.fi/cost219/cosb229b.HTML) has been collated from available guidelines and recommendations. The material has been ordered in three groups:

- Issues to consider in order to fulfil 'Design for All' requirements
- Issues to consider in order to make the product adaptable in case 'Design for All' is not possible
- Issues to consider if even adaptations to assistive devices are not possible.

The general requirements cover 26 items from 'finding the terminal' to 'user dialogue of telematic services'. The listing is mainly meant to give a general overview of requirements without yet considering the specific device which is being developed. For instance, if we choose item 14, 'read screen' we find the requirements and suggestions as shown below.

### 13.4.1 Read screen/display

**Design for All**: Provide large letters (e.g., 9 mm on LCD) for textual/numeric output. Normal lower case text is easier to read than text in only capital letters. Do not underline. Use good contrast. Avoid red/green and yellow/blue combinations. Use, for example, black on light yellow or black on white. Avoid saturated primary colours (red, green, blue) near to each other. Height of characters should not be less than 4 mm. Font sans-serif is considered easy to read. 16 point bold on screen can be read by many people with visual disabilities. Provide choice to select font size. Provide selection of colours, brightness and contrast for screen operations. Background colour should be selectable. Provide magnification of screen object and cursor for screen operations. Give control information on screen in optional ways (e.g., icons, pull down menus with text). Icons should also have text labels. Use standardised symbols. Keep symbols and icons simple without too many small details. Mobile telephones should have text capability.

**Adapted**: Provide speech synthesised output for textual/numeric output. User preference for voice output could be coded on the card (CEN prEN 1332-4).

**Special**: Text telephones should provide possibility to use voice during text call. Videophones should provide possibility for voice and text transfer. Standard interfaces for optional output devices.
www.stakes.fi/cost219/cosb229b.HTML

It is clear that these general guidelines are not as detailed as the specific guidelines that are found for the most common devices.

## 13.5 SPECIFIC COST 219bis GUIDELINES

COST219 bis has developed a series of specific guidelines, which are now open for comments (www.stakes.fi/cost219/cosb235.htm):

- Accessibility Requirements for New Telecommunication Equipment (e.g., smart telephone, palmtop, sub-notebook, set-top box—see Figure 13.2)
- Producing Web Pages that Everyone can Access
- Design Guidelines on Smart Homes
- Guidelines-Booklet on Mobile Telephones
- New Ways of Using Video Telephony
- Pay telephones with immediate public access
- Text telephony for deaf, hearing-impaired, deaf-blind and speech-impaired people.

These guidelines are generic, in the sense that they highlight the problems encountered by people with reduced abilities, indicating overall solutions without always going into specific details of how to implement the features in practice—which is left up to the manufacturers. At the same time, the guidelines are also quite detailed in some instances (for example, size of keys, height of pay telephone off the ground, etc.).

The target audience ranges from decision-makers, telecommunication designers and information service providers through to local authorities, user groups and the customers themselves. For example, the guidelines on 'Producing Web Pages Everyone can Access' are targeted at Web page designers, information service providers, user groups and any decision-maker who is planning to set up a Web site. On the other hand, the 'Design Guidelines on Smart Homes' are targeted more towards local authorities, emergency and care service providers and the user groups.

The guideline summary page (www.stakes.fi/cost219/cosb235.htm) also gives links to all other stable guidelines that have been found on the Web.

Let us, as an example, take some excerpts from the first specific guideline (www.stakes.fi/cost219/smartphones.htm;
and see also www.rnib.org.uk/wedo/research/sru/phones.htm).

**Figure 13.2** New telecommunication equipment

### 13.5.1 Keys (keyboard, keypad, buttons)

Due to the limited dimensions of these types of equipment, their keyboards and keypads are often intended to be used with one or two fingers at most. The general goals in designing key shape are the following:

- The finger should locate the key without hitting other keys
- Hitting other fingers should be avoided if keys have to be used for multi-finger typing
- The distribution of pressure should indicate the location of the finger on the key
- The force of pressing the key should be distributed to the proper part of the finger.

A selection of general guidelines concerning the design of keys include the following:

- Use clearly recognisable shape. Choice of round versus square keys may be a cultural variable (e.g., round or oval keys are more popular in Germany than in the US)
- Top of keys concave or at least flat
- Provide tactile identifiers on keys:
  Raised dots on number 5 in numeric keypads
  Raised bars on F and J keys on typing alphanumeric keyboards
  Consider printing or engraving key labels to give texture
  Provide for clear tactile separation of keys when arranged in groups
- Good contrast between keys and body of the equipment
- Sharp contrast for labels on keys using readable big fonts.
  Minimum 10–12 point for hand-held equipment, 14 point for devices to which the user reaches (device lying on a table)
- Automatic repetition of key function should be avoided or user configurable
- Provide clear tactile feedback (snap action).

### 13.5.2 Design guidelines for interaction methods

There are various design guidelines regarding general issues in user interaction techniques and user interface design. You can find some useful references at the end of the guideline in question.

A selection of very general guidelines particularly suited to user interfaces of lightweight telematic terminals are given below:

- Avoid burdening user memory
- Use direct selection techniques for common functions
- Avoid nesting menus or dialogues with more than 3 levels
- Allow the user to redo previous selections
- Provide cancel function for operation in progress
- Provide undo function whenever feasible
- Provide on-line help and short explanation of selected commands whenever possible
- Provide for selectable fast/easy interaction for skilled/unskilled users.

The practical implementation of accessibility, 'Design for All' or 'Universal Design' requires that not only the designer but all others in the company actively embrace the 'Design for All' philosophy. This has been treated in the *Handbook on Inclusive Design* produced by the INCLUDE project. The issues have also been developed in the industry awareness work in COST219bis.

## 13.6 HANDBOOK ON INCLUSIVE DESIGN

A Handbook for designers with concrete checkpoints to follow for inclusive design has been produced and is also available on the Web at www.stakes.fi/include/handbook.htm It begins with a chapter on common claims and challenges in contemporary product development. It is shown that arguments like

- Disabled and older people are not our target markets
- Optimising profitability is optimising time to market, or
- It is too expensive and time-consuming to involve end-users

are either false or contradictory to the 'Design for All' principle. The Handbook continues with the following chapters:

- The product development process from the point of view of different departments; checkpoints for inclusive design are presented for each phase throughout the process
- Management of human-centred, usable, accessible outcome
- Inclusive defining of target markets
- Creating inclusive product/ service concepts
- Design
- Testing and refinement
- Exploitation, commercialisation and after sales
- Summary of checkpoints for inclusive design
- Links to guidelines for inclusive design, both general and specific
- Case studies.

The Handbook shows how important it is to have a genuine commitment to inclusive design in an organisation and how to find the information needed for the implementation of 'Design for All'.

## 13.7 WHAT RESEARCH SHOULD DO

As well as a commitment to 'Design for All' on the part of individual organisations, it is very important to build up and draw attention to accessibility and inclusive design in the day-to-day management of public research programmes. The managers of research programmes, as well as the staff, should:

- be aware of the needs of disabled people and older people
- check that these needs are taken into account in all relevant research.

If the project officers in the various research programmes do not really support these aims, then neither will the projects. Staff will not really support the

'Design for All' concept if their superiors do not. Therefore, research managers need to be convinced of the importance of accessibility and inclusive design. This can be done by showing that it follows from the anti-discrimination provisions in the Amsterdam Treaty and that good design also makes products and services more competitive on the market.

Programme managers need to convince their staff to adopt a 'Design for All' philosophy. They should clearly state accessibility and 'Design for All' as requirements for all relevant projects, and these requirements should be included within the evaluation criteria for project proposals. During the project's negotiation phase, project officers should ensure that accessibility requirements and 'Design for All' issues are considered when appropriate in the research and development work and that the consortium is aware of the information and guidelines that are available in this field. Project proposals should identify:

- The human-centred good practice design process activities—understanding and identifying context of use, specifying user and organisational requirements, producing prototypes and evaluating designs according to user criteria
- Procedures for integrating these activities with other system development activities, e.g., analysis, design, testing
- The individuals and the organisation(s) responsible for the human-centred design activities and thc range of skills and viewpoints they provide
- Effective procedures for establishing feedback and communication on human-centred good practice design activities as they affect other design activities and methods for documenting these activities
- Appropriate milestones for human-centred activities integrated into the overall design and development process
- Suitable time-scales to allow feedback and possible design changes to be incorporated in the project schedule.

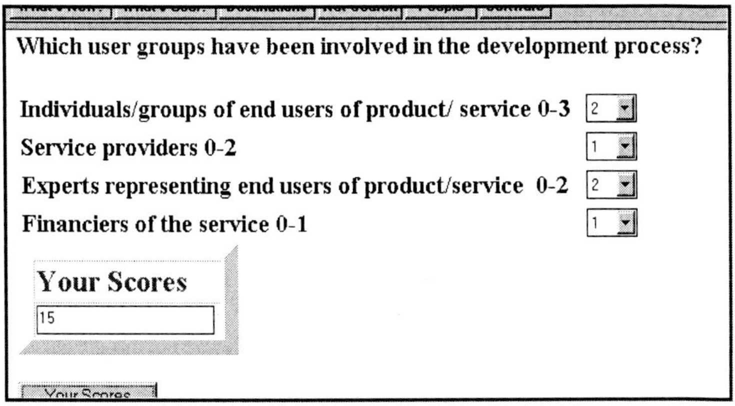

Which user groups have been involved in the development process?

Individuals/groups of end users of product/ service 0-3 — 2
Service providers 0-2 — 1
Experts representing end users of product/service 0-2 — 2
Financiers of the service 0-1 — 1

Your Scores
15

**Figure 13.3** *INCLUDE checklist*

INCLUDE has produced a simple one-page checklist (Figure 13.3) that can be used in order to check how well disabled users and elderly users have been considered when appropriate. This should be considered in the procedures from the very beginning. The checklist is also available on the INCLUDE Web site for developers to perform a quick self-evaluation of the inclusiveness of their design process (www.stakes.fi/include/incd431.html). To ensure good practice, projects should, of course, also follow and implement the ISO 13407 standard on *Human-Centred Design Processes for Interactive Systems.*

## 13.8 CONCLUSION

The information society can become accessible if all involved in developing society or developing and implementing services and manufacturing products realise that all citizens do not have the same abilities or preferences. Following the guidance provided by COST 219bis and the INCLUDE project will help to achieve this aim.

CHAPTER FOURTEEN

# Public Access Terminals

John Gill

## 14.1 INTRODUCTION

To fully participate in society, individuals will need to be able to use self-service terminals. Many government departments may have plans for using public access terminals for providing information, collecting taxes, granting licences, administering regulations, paying grants and benefits, collecting and analysing statistics, and procuring goods and services. Some of these services may also be available through direct contact with a human, but there may be an additional charge for using this facility.

To make significant progress in the accessibility of public access terminals by disabled and older people will require terminal manufacturers and service providers to adopt a 'design for all' policy. In addition there will need to be agreement to standardise a number of aspects of the user interface of such terminals.

## 14.2 THE SCIENTIFIC BASIS

Occasionally recommendations for standards for the human–computer interface are based on rigorous scientific research with a cross-section of potential users. Unfortunately this is expensive and therefore is often omitted or done with a very small sample of potential users. This is not necessarily a serious problem in that often the recommendations could be described as 'common sense', but it can lead to standards which are later proved to be fallacious.

A common method for deriving a standard is to quote the contents of another standard without verifying whether the detailed recommendations are appropriate. Sometimes recommendations have been what is thought best within certain technological limitations but the recommendations have not been altered when the technology changed.

For instance, the typeface for subtitling on analogue television was limited to a mosaic typeface. However, with digital television the limitations are fewer but programmes made for analogue television must be able to be broadcast on digital television without re-authoring the subtitles. In this case, the existing recommendations for typefaces for people with low vision were examined but were found to have obvious flaws. For instance, they often recommended the use of the Arial typeface when representing numerals (Figure 14.1). However, these could be easily confused and more open characters would be more discriminable (Figure 14.2; see www.tiresias.org/fonts).

689

**Figure 14.1** Arial typeface

689

**Figure 14.2** Tiresias screenfont

Other problems can occur with the adoption of new assistive technology. For instance, scooters used by physically disabled persons are significantly different in height from conventional wheelchairs, which means that recommendations for the height of keypads and screens on public terminals may no longer be valid.

In other areas guidelines have been constrained by external factors. For instance, most people with low vision would benefit from a high level of illumination in the immediate vicinity of an outdoor cash dispenser. However the bank staff responsible for security do not want such illumination since it makes life easier for muggers; the customer withdrawing money would be looking from a bright area to a much darker area and might not see someone waiting to rob them. Therefore the recommendations on illumination levels have had to be compromised between what is beneficial for low vision persons and what is acceptable to the bank. However well-lit areas, not just at the cash dispenser, would make the environment safer for everybody.

## 14.3 GENERAL DESIGN CONSIDERATIONS

Different groups of people with a disability have a variety of problems with cash dispensers (Table 14.1). For many disabled and older users, the most important aspect is consistency in the user interface of public terminals; this is particularly important for visually, intellectually and cognitively impaired users. A prime example of this is the lack of a single standard relating to the layout of numeric keypads. With public terminals, the user may only use it occasionally and has probably been provided with minimal training in the use of the terminal. What is 'logical' to the average user may be different from what is 'logical' to the designer, so it is essential to test any new user interface with a cross-section of potential users (including disabled and older people).

To select a preferred interface such as audio instructions or large characters on the screen, the user could simply press a button or otherwise select from a menu on the screen; this is likely to increase the time taken to undertake a transaction if there are more than a few options. Another possibility is to store the user's preferences on a central computer and implement them as soon as the PIN (personal identification number) has been entered.

For card-operated terminals, it is possible to store the information on the user's card (the coding of user requirements is specified in the European standard EN1332–4), and this is in many ways more desirable than storing private information about a user on a central database. With a magnetic stripe card there is very limited spare capacity for storing this information (but this method has been used for storing the user's preference for displayed language), although a smart card (containing an electronic chip) has fewer restrictions on storage capacity so appears to be ideal for this purpose.

**Table 14.1** Problems with a cash dispenser for different groups of people with disabilities

| | Wheelchair user | Cannot walk without aid | Cannot use fingers | Cannot use one arm | Reduced strength | Reduced co-ordination | Dyslexia | Intellectually impaired | Blind | Low vision |
|---|---|---|---|---|---|---|---|---|---|---|
| Locate terminal | Few | Few | Few | Few | Few | Few | Few | Some ● | Many ● | Some ● |
| Access to terminal | Many ● | Some ● | Few | Few | Few | Few | Few | Few | Few | Few |
| Read instructions | Few | Few | Few | Few | Few | Few | Some ● | Many ● | Many ● | Many ● |
| Insert card | Many ● | Some ● | Many ● | Some ● | Some ● | Many ● | Few | Some ● | Some ● | Some ● |
| Read screen | Many ● | Few | Few | Few | Few | Few | Some ● | Some ● | Many ● | Many ● |
| Use keyboard | Many ● | Few | Many | Few | Few | Some | Few | Some | Some ● | Some ● |
| Use touchscreen | Many ● | Few | Many | Few | Few | Some | Some ● | Some | Many ● | Some ● |
| Retrieve money | Some | Some ● | Many | Some | Few | Some | Few | Few | Some | Few |
| Read receipt | Few | Few | Some | Few | Few | Few | Some | Some | Many | Many ● |
| Retrieve card | Many ● | Some ● | Many ● | Some ● | Some ● | Some ● | Few | Few | Some ● | Few |

| Few problems | Some problems | **Many problems** |
|---|---|---|

● Technology available to alleviate the problem

Many disabled users would like to be able to select and store their preferred interface whenever they use their card at a public access terminal. It is essential that information is stored on a card only with the consent of the user.

### 14.3.1 Locate terminal

For a blind person, it can be difficult to find the terminal if they are not very familiar with the environment. One possibility is to use a contactless smart card, carried by the blind person, to trigger an audible signal from the terminal at a distance of a few metres (Figure 14.3).

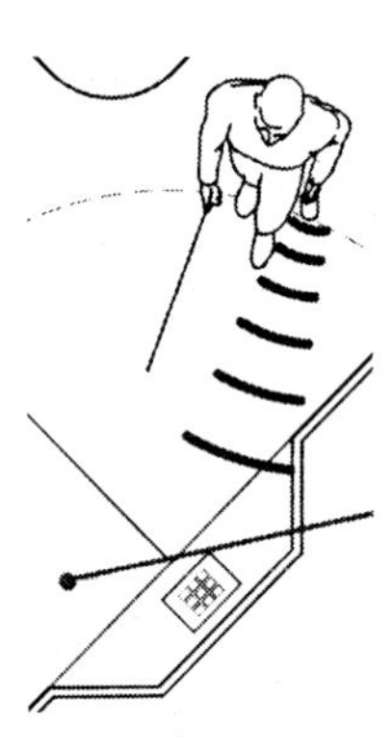

**Figure 14.3** Audible location signal for a terminal

### 14.3.2 Instructions

Instructions should be written in simple clear language and presented at eye height in at least 16 point bold characters, preferably in white or yellow on a dark matt background. It is important that the instructions are not worn away with use of the terminal (or they should be replaced periodically). It is often useful to number the instructions and then associate by number with the physical parts of the interface (e.g., card reader) as well as showing the number on the visual display. Ideally the user should be able to choose the language; frequently this is only viable if the instructions are displayed on the screen. It would be preferable if the user's card stored their preferred language and that the terminal automatically switches to this as soon as the card is inserted.

Public access terminals can incorporate audio prompts in the form of 'beeps' to indicate an action. It is recommended that new equipment should provide a more sophisticated solution of using audio leadthrough in the form of a verbal set of instructions. Audio leadthrough can assist people with visual or cognitive impairments (and first-time users). Sentences should be concise and simple in structure and only natural vocabulary should be used. Informative messages which advise the user of the progress of the transaction and inform the user when or how to perform a step in the transaction should be clear and to the point, and provide confirmation of task completion. Message content should be chosen very carefully since a message that might be acceptable to the users for the first few times they hear it may become unacceptable when they hear it for the hundredth time. Many users with impaired hearing, who can only hear lower frequencies, can more easily hear a male voice than a female one.

If audio output is used to provide private information to the user, then it should be through a telephone handset located at the terminal or through a headset connected through a standard mini-jack to the terminal; however, it is essential that the position of the jack socket is standardised (Figure 14.4). If a handset is provided, inductive coupling and amplification should also be incorporated.

Braille instructions on outdoor terminals have limited value in cold weather since tactual sensitivity is dramatically reduced with decreasing temperature. The estimated number of Braille readers in Europe is less than 200,000, so although useful for some blind users, Braille is not a total solution for visually impaired users.

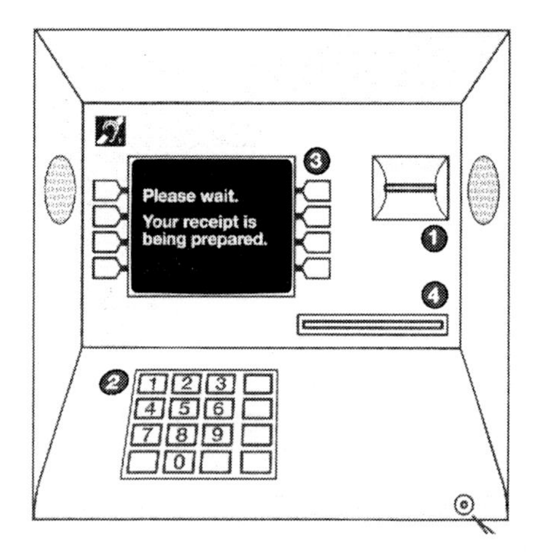


**Figure 14.4** An ATM with mini-jack for audio output

### 14.3.3 Card insertion

For a blind person, there is a problem in selecting the right card from their wallet; unfortunately there is no standard method for tactually marking cards to indicate the issuer or their use. This problem will be exacerbated with the increasing use of cards which feel the same.

For the naïve user, it is often far from obvious where to insert the card. A flashing light around the card entry slot has been found beneficial. For those with hand tremor, it is useful if the entrance to the card reader acts as a funnel to guide the card in correctly (Figure 14.5).

Blind persons, and many older persons, have problems in inserting the card in the correct orientation; this is a particular problem on cards which are not embossed. However there is a European standard (EN1332–2) for an orientation notch in the card (Figure 14.6).

For many wheelchair users, such as those with arthritis, it is not just a problem of reaching the card reader but still having any useful grip as the arm is raised above the horizontal.

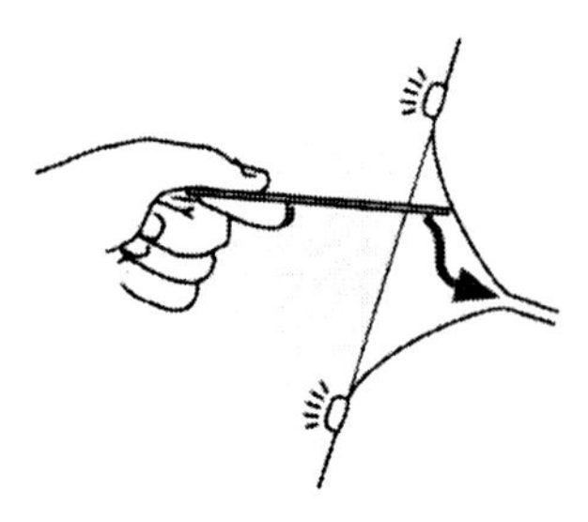

**Figure 14.5** A funnel-shaped entrance facilitates card insertion

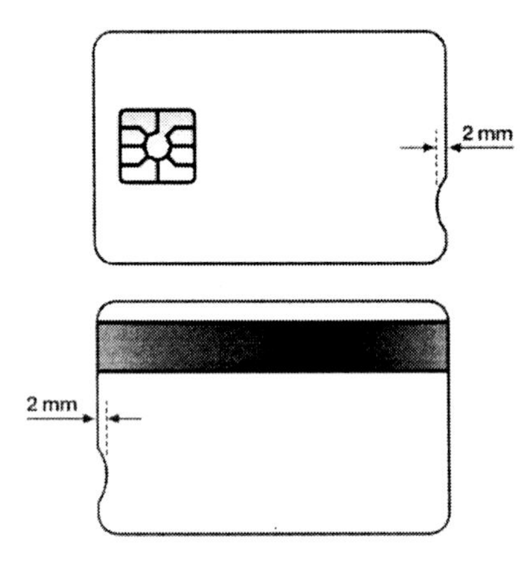


**Figure 14.6** An orientation notch according to EN 1332–2

The lowest height of any operable part of the user interface should not be less than 0.7 metre. Ideally the terminal, or user controls, should be adjustable in height; although this is done on some drive-in cash dispensers, it does significantly increase the cost of the terminal. However, the problems of accessing the card reader are greatly alleviated if contactless smart cards are used; for this type of application they typically have an operating range of 10 to 20 cm.

### 14.3.4 Reading the screen

People who wear bifocals find it difficult to read the screen of most public access terminals since neither lens is in focus at the distance between their eyes and the screen. In addition many people leave their spectacles in the car or do not wear them in public. So the number of people who have problems in reading the screen is much more than those considered 'blind' or 'low vision', who constitute about 1.5% of the population.

People with low vision should not be prevented from positioning their faces close to the screen. However it is possible to increase the size of the characters on the screen for individual customers who require this facility. This can be done by selecting this option from a menu or preferably by storing this information on the customer's card. With touchscreen systems, it could be arranged that holding one's finger in the top left corner for at least two seconds indicates that one would like double size characters on the screen (Figure 14.7).

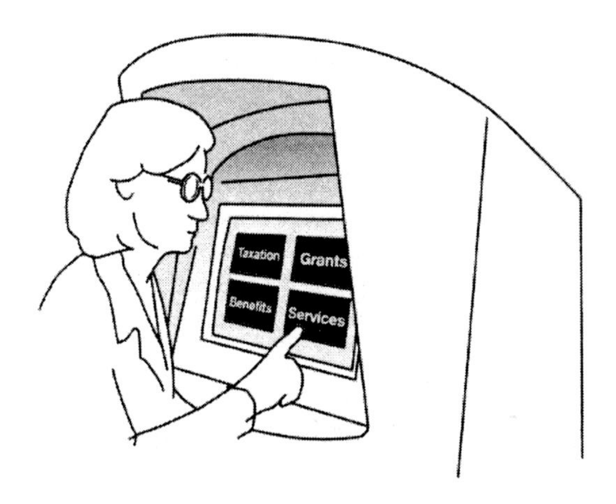


**Figure 14.7** A touchscreen with large characters

Total colour blindness is rare (less than 0.0025% of the population) but problems with discriminating red and green are common (over 6% of the male population).

Moving text on a screen can be very difficult to read for someone with even a mild sight impairment, so it should be avoided whenever possible.

Digitally stored speech can give very good audio quality but it is effectively limited to pre-stored messages. Full vocabulary synthetic speech is often difficult to understand for the naïve user, particularly if they have a hearing impairment. Non-confidential information can be output on a loudspeaker, but the volume should be a function of the current ambient noise level; this is less of a problem with handsets or headphones. If there is an inductive loop for hearing aid users, there should be a clear visual indication (Figure 14.8) that this is the case (note that not all hearing aids have facilities for loop connection).

**Figure 14.8** The sign to indicate inductive coupling for hearing aids

One technological possibility would be for a disabled user to have a hand-control unit with an infra-red link to the terminal. This would require all terminals to use the same interface protocol and care would be needed to ensure confidentiality of sensitive information (Figure 14.9).

**Figure 14.9** Hand-control unit with infra-red link to the terminal

Sunlight can degrade the viewability of the display for all users. The site should be one where direct or reflected sunlight or other glare is prevented from striking the visual display. The display should be viewable from the eye level of a person sitting in a wheelchair or on a scooter; this results in problems in specifying absolute heights since the dimensions of scooters vary considerably. One solution, albeit expensive, is to have the user interface on the terminal be adjustable in height; this is particularly important if the terminal is likely to be used by children as well as adults (as in a post office).

The conflicting requirements from tall pedestrian users and short wheelchair users can lead to a significant group of users having parallax problems when lining up the function keys with the displayed option. Lines on the user interface leading from the key to the surface of the display can alleviate this problem (Figure 14.10).

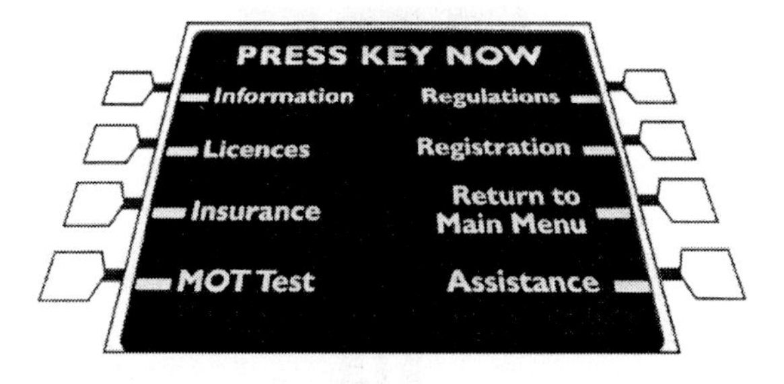


**Figure 14.10** Parallax problems between function keys and text on the screen

Displayed text should use simple, large, bold fonts in upper and lower case characters. Displayed messages should be simple in sentence structure, use natural language and any graphical symbols (such as icons) should be accompanied by text.

### 14.3.5 Keyboards

Personal identification numbers (PINs) are a particular problem for many dyslexic and intellectually impaired people. Both these groups would find it advantageous to have the option of using a biometric method for identification (e.g., fingerprint).

With biometric methods of identification, it is essential that users have a choice between the biometric method and some other method (e.g., PIN). This is because for many biometric systems there is some group of disabled people who cannot use it (e.g., fingerprint identification requires the user to have fingers), the exception being facial recognition.

The user's PIN should not be displayed, printed or broadcast by any means. However, it would be useful to have both an audible feedback and a visual one (e.g., an X on the screen) to show that a digit has been input. Many people with even slight memory problems find it difficult to remember and input their PIN quickly, so it would be helpful to allow a generous amount of time before they are timed out.

Information which is sensitive and private to the cardholder should not be visible to any other person; screen filters improve privacy but often at the expense of visual quality. However, the user may wish to display information with a large character size, in which case they should be made aware of the privacy problem.

Standard layout of keypads is essential for visually disabled people and highly desirable for other users. To help blind persons, there should be a single raised dot on the number five key (Figure 14.11). However this does not solve the problem of there being two common layouts for the numeric keys (i.e., the telephone and the calculator layouts); it is recommended that the telephone layout is used exclusively on public access terminals; for example, see www.tiresias.org/phoneability/telephones

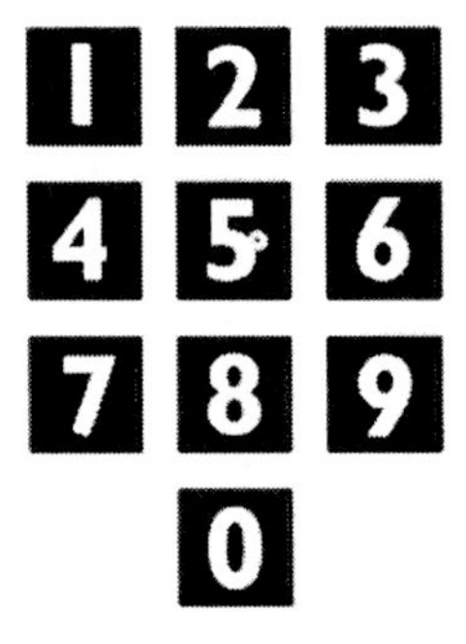


**Figure 14.11** Telephone keypad layout with an embossed dot on the number 5

All keys or buttons should be tactually discernible; keys should be raised or recessed by a minimum of 2 mm. The edges of the keys should be at least 2.5 mm apart. Function keys should be clearly separated from the numeric keys.

Visual markings on the keys should be characters of at least 4 mm high and should have good contrast with the colour of the key (e.g., white characters on matt black keys). Where text keys are colour-coded, they should be coloured as in Table 14.2.

**Table 14.2** Colour-coded keys

| Key meaning | Colour |
|---|---|
| Enter or proceed | Green |
| Clear or correct | Yellow |
| Cancel | Red |

Colour should not be the only distinguishing feature between keys, since red/green colour blindness is not uncommon; if possible, the keys should have different shapes and be marked with symbols.

Ideally keys should be internally illuminated when the terminal is waiting for input from that keypad. There should be some form of feedback on key input (e.g., a beep and/or tactual indication). Tactile feedback can be provided by a gradual increase in the force, followed by a sharp decrease in the force required to actuate the key, and a subsequent increase in force beyond this point for cushioning.

Many older people and those with a cognitive impairment do not like to be rushed or to think that they are likely to be 'timed out' by the machine, so it is necessary to allow for such people to use the terminal at their own pace. This requirement could be stored on the user's card.

Speech input for commands is an option in some situations. If this is adopted, then the user should have the choice of keyboard or speech input. It is likely that speech input would be preferred by people without hands and those with an intellectual impairment but the keyboard is easier for those with a speech impediment.

### 14.3.6 Touchscreens

To help older people and those with hand tremors, key fields should be as large as possible and separated by a 'dead area'. There should be high contrast between touch areas, text and background colour. Avoid using a pretty picture as background—it is a menace to anyone with poor vision or someone reading the screen under difficult conditions, e.g., in bright sunlight (Figure 14.12).

For blind users, one possibility is to arrange that holding one's finger in a specified corner of the screen for at least two seconds initiates speech output (note that this must be a different corner than the one used to request large characters on the display), or tapping twice in the corner (Figure 14.13). Another method would be to store this requirement on the user's card.

**Figure 14.12** Touchscreen input

**Figure 14.13** Selecting special outputs on a touchscreen

Touchscreens can either be triggered by insertion or withdrawal of the fingertip. With the latter system, it is technically possible for the user to pass their fingertip over the screen and get speech output describing the active area they are touching at the time. Then the system is only triggered by withdrawing the fingertip from over an active area.

### 14.3.7 Money retrieval

Cash, receipt, or any other document issued from the terminal for withdrawal by the user should protrude at least 3 cm beyond the slot surround.

For someone with poor manual dexterity, such as with arthritis, taking a card from a terminal and then taking the money may be difficult to do in the allowed time. Increasing the time for everybody increases the security risk. However, it would be possible to let users decide if they want more time than the standard time permitted and to store this requirement on their card.

Security at cash dispensers is a major concern for many older people and is often given as a reason for not using such terminals. Therefore anything which improves the user's perception of safety is to be welcomed.

### 14.3.8 Receipts

To aid visually impaired users, receipts should have a minimum font size of 12 point with a sans serif typeface with upper and lower case text, but 16 point would be preferable if space permits. It is important that the print has good contrast on opaque paper with a minimum of background pattern. A common complaint is poor print quality on receipts which is often a result of the printer ribbon not being replaced regularly.

### 14.3.9 Card retrieval

Many people with arthritis have difficulty in gripping and pulling the card from the reader, particularly when the arm is extended above the horizontal. The card should protrude at least 2 cm from the slot surround. Therefore it is recommended that the force necessary for the user to retrieve the card from the terminal should be not any greater than that needed to stop the card from falling out of the reader.

### 14.3.10 Training

All too often it is assumed that users will learn to operate a self-service terminal just by reading the brief instructions attached to the terminal. However, the instructions must be comprehensible to all users and not just the system's designer. For instance, the instruction "Enter your PIN" is not meaningful to many older persons; an alternative might be "Please key in your personal number now".

Service providers have proved very reluctant to provide users with training in the use of their terminals. This has resulted in many older persons not using services because they could not operate the terminal.

## 14.4 CONCLUSION

The increasing use of terminals intended for use by the general public makes it essential that the needs of disabled and older people are taken into account when the systems are being designed.[1] Many features which are essential for disabled people are advantageous for all users.

---

[1] Many of these guidelines have been developed as part of the work of COST 219bis (www.cost219.org) and the INCLUDE project (www.stakes.fi/include) and are further documented on www.tiresias.org/pats

## CHAPTER FIFTEEN

# Accessible Systems for Transport

Colette Nicolle and Gary Burnett

## 15.1 INTRODUCTION

Maintaining mobility is a primary objective of people who are older or disabled. Travelling by car or public transport enables them to partake in work or leisure activities outside the home, without a reliance on others to do so, thus sustaining independent living and quality of life. However, transport vehicles, systems and services are not always designed with their needs and abilities taken into account and so older and disabled people are often less likely to travel than other citizens. This may be because the transport has not been designed for all, or because the traveller does not have enough clear information to know whether or not the transport meets specific needs.

New technologies in the travelling environment, called intelligent transport systems (ITS), or advanced transport telematics (ATT), can assist travellers, especially those who are older or disabled, both before and during a journey. Some ITS may provide information to the traveller, some may provide warnings in hazardous conditions and some may assist a driver in controlling the vehicle. Emergency call systems and route guidance and navigation systems may increase safety and restore confidence for older or disabled drivers, provided that the controls and displays are designed with their functional impairments and their requirements in mind.

Travellers who are older or disabled may be the most likely to benefit from these systems but in reality may have difficulties in taking advantage of them due to the very same limitations. Thus, as vehicles become faster, our roads busier and the able-bodied traveller seemingly more efficient, older or disabled travellers may find themselves lagging further behind and with potentially increased safety risks—unless ITS are designed with their functional impairments and their requirements in mind. It is therefore important that there are specific guidelines to ensure that these systems are both easy to use and contain relevant information to meet the specific needs of all travellers, in particular those who are older or disabled. Inclusive guidelines which benefit older or disabled travellers will also benefit a much wider sector of the population. For example, wheelchair-accessible, low-floor buses are also helpful for parents with small children or when carrying heavy shopping.

In the context of travelling in the UK, the 1995 *Disability Discrimination Act* (DDA at www.disability.gov.uk/dda/index.html) enables the Government to make regulations to require that all new public transport vehicles are accessible, thus ensuring that disabled people can get on and off transport vehicles and travel in safety and reasonable comfort. Part III of the DDA on *Access to Goods, Facilities and Services* is also very relevant, since transport must be accessible in order for older and disabled people to take advantage of the service. Even though the vehicle may be accessible, however, travellers need to know that it is accessible—otherwise they may never feel confident even to begin their journey. So the provision of information is equally important.

Although many product designers and developers of ITS realise that it is essential to consider the needs of older and disabled people, they may not be aware of their particular requirements, how to go about identifying them, and then how to ensure that all the users' main problems and concerns are considered in the design process. Designers would, therefore, benefit from various types of guidelines:

- Prescriptive guidelines on how to design the system's interface so that all travellers are able to use it more easily (e.g., to make the system flexible enough and easy to adapt to individual user requirements)
- Prescriptive guidelines on what type of specific information is needed by older or disabled travellers (e.g., to include accessibility issues in trip-planning systems, such as the number of stairs to climb at train stations)
- Process guidelines which provide methods, tools and protocols to identify the users' requirements and to evaluate systems for ease-of-use by people who are older or disabled.

Based on the work of the TELSCAN project, described below, this chapter provides an overview of such prescriptive and process guidelines for systems for travellers, describing how they were developed, and making recommendations for the future to ensure that such guidelines are both used and useful.

## 15.2 BACKGROUND

### 15.2.1 European research projects

The TELAID (Telematic Applications for the Integration of Drivers with Special Needs) project was part of the European Union's DRIVE II Programme and ran from 1992 to 1995. TELAID investigated the requirements of people with disabilities who need the use of car adaptations whilst driving and what implications these might have on the use of advanced technologies. The DRIVE II EDDIT (Elderly and Disabled Drivers Information Telematics) project focused on the elderly driver's use of ITS, using conventional vehicle controls but illustrating the users' normal, gradual deterioration in perceptual and performance abilities. TELAID then joined with EDDIT to form the TELSCAN project, which ran from 1996 to 1999. Part of the European Union's Telematic Applications Programme, TELSCAN (TELematic Standards and Co-ordination of ATT systems in relatioN to elderly and disabled travellers) broadened the research to include the needs of older and disabled people in the development and application of ITS, whether it be as drivers using their own cars, or as travellers in various modes of public transport.

Out of TELSCAN came different types of guidelines which can be used in both the design and evaluation of systems for travellers:

- Prescriptive guidelines in the form of a handbook and database of interface design guidelines. These ensure that systems are designed to be usable by older people and people with disabilities.
- Prescriptive guidelines in the form of a travel information checklist. This ensures that travel information systems provide older and disabled people with all the information they need before setting out and whilst on a journey.
- Process guidelines in the form of an assessment methodology. This provides guidance to ITS designers to include older and disabled users in their evaluation plans.

### 15.2.2 Identification of requirements

The underpinning of any guideline is the identification of the users' requirements—only then can systems be designed with these needs in mind. During the TELAID and TELSCAN projects, requirements were identified for travellers having a wide variety of impairments and using a range of public and private transport modes. Four distinct types of user requirement were identified and the following diagram (Figure 15.1) demonstrates how they have been considered in the development of guidelines for ITS. The illustration also emphasises the relationship between the primary outputs of the TELSCAN project—*the Design Guidelines Handbook*, the *Travel Information Checklist*, and the *Assessment Methodology*. It is not enough that the system interface is designed and evaluated with the user's needs in mind—it must also contain relevant information for different user groups. Following such guidance will lead to more usable intelligent transport systems for all.

#### *15.2.2.1 Interface requirements*

People who are older or disabled may experience difficulties with various aspects of the system's interface. For instance, a system which utilises a visual display containing text may be a problem for those with poor sight and/or language skills. Consequently, travellers' interface requirements have contributed to two important products of the TELSCAN project:

- *Design Guidelines Handbook* (Nicolle and Burnett, editors, 1999). This provides guidelines on interface design to ensure that ITS are easy to use by older and disabled travellers
- *Assessment Methodology* (Marin-Lamellet *et al.*, 1999). This document outlines the methods, tools and protocols that are required so that older and disabled people can be adequately included in the assessment process of ITS.

#### *15.2.2.2 Information requirements*

Systems for travellers should include information which an older or disabled person needs in order to make a journey safely and in comfort. For example, travellers

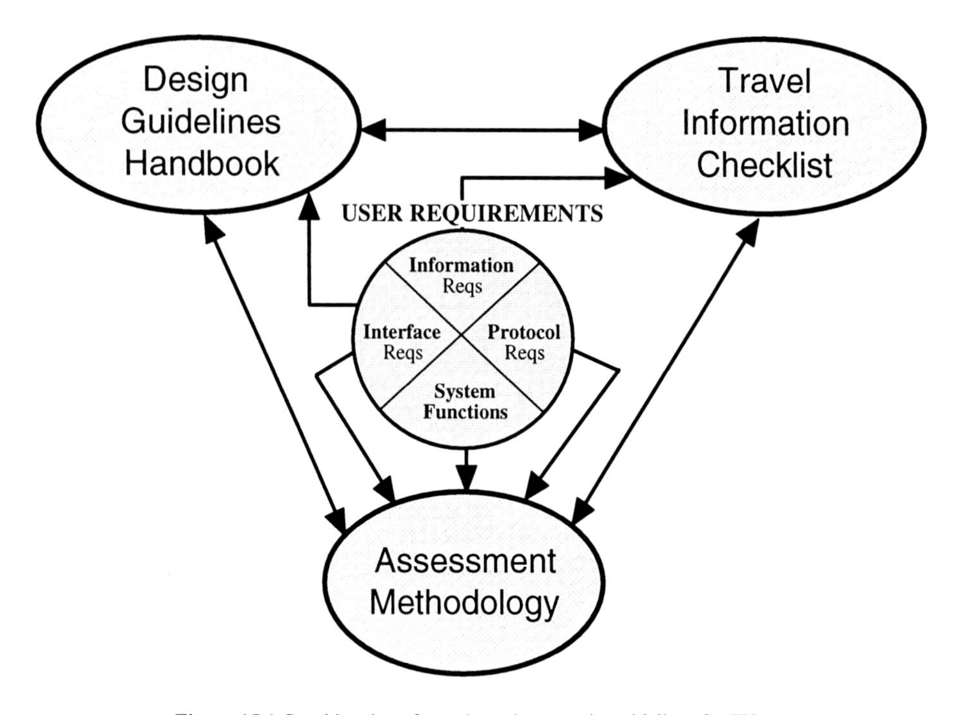


**Figure 15.1** Consideration of users' requirements in guidelines for ITS

may need to know:

- which buses have a low floor for easier access
- whether there are reduced fares for people who are retired or disabled
- how long it will take to walk from one platform to another
- if there are any stairs to climb.

These information requirements have been developed within TELSCAN as the *Travel Information Checklist* (Veenbaas, 2000) to give designers this sort of advice.

### *15.2.2.3 Protocol requirements*

The users' requirements, as well as our experiences in evaluation activities, have identified certain protocols that should be followed when including older and disabled people in the design and evaluation process. These form part of the TELSCAN project's *Assessment Methodology*. Various methods and tools either may not be appropriate for certain people or may require specialist knowledge in their use; for example, guidance is provided on the use of questionnaires or focus group interviews with older or disabled people. The methodology draws advice on data collection methods from USER*fit* (Poulson *et al.*, 1996), described in this volume.

### *15.2.2.4 System function requirements*

TELSCAN has identified the numerous difficulties experienced by elderly and disabled people when travelling via different modes of transport (Nicolle *et al.*, 1997; Nicolle *et al.*, 1998). Such problems can be alleviated via the functionality offered by ITS; for instance, a person using a wheelchair might benefit from increased knowledge to aid in trip-planning, such as whether there are any stairs at the destination station.

These requirements are also reflected in TELSCAN's *Assessment Methodology*, which helps evaluators identify which different user groups are *most likely* to benefit from the provision of certain high-level functionality within a system, although obviously the functionality of ITS can be considered important for all travellers.

It is clear from our analysis of users' needs that recommendations may point the way towards a new type of system. More often, though, they point towards prescriptive guidelines covering particular information (section 15.4) or interface design requirements (section 15.3) for an existing system. Such guidelines are presented below. These are followed by process guidelines for assessing the system (in section 15.5) so that it meets those users' requirements and thus will be more usable by people who are older or disabled.

## 15.3 GUIDELINES FOR INTERFACE DESIGN

Interface design guidelines are needed to ensure that older and disabled people are not excluded from using and benefiting from intelligent transport systems. To illustrate, if a system includes controls which are too close together, this may cause particular difficulties for older travellers with co-ordination or dexterity limitations, as well as those with severe upper-limb impairments. This means that a system should follow inclusive design principles, or should at least be easily adaptable to meet specific requirements.

The *TELSCAN Code of Good Practice and Handbook of Design Guidelines for Usability of Systems by Elderly and Disabled Travellers* was developed as a guide for use by designers during both the design and evaluation process. A distinction is necessary here between the designers as users of the *Handbook*, and older and disabled people as potential users of the ITS which are being developed. The requirements of both types of users have to be taken into account: the guidelines handbook also needs to be usable (and used) by designers in order to ensure that the systems are usable by older and disabled travellers.

The following diagram (Figure 15.2) sets out the six basic parts of the *Handbook* and their inter-relationships. The inverted triangle demonstrates the general progression of the guidelines from general to specific. Users of the *Handbook* are reminded that, no matter what type of system is being developed, they should be familiar with the general guidance, especially in Parts 2 and 3, as these relate to all systems and will constitute good 'design for all'.

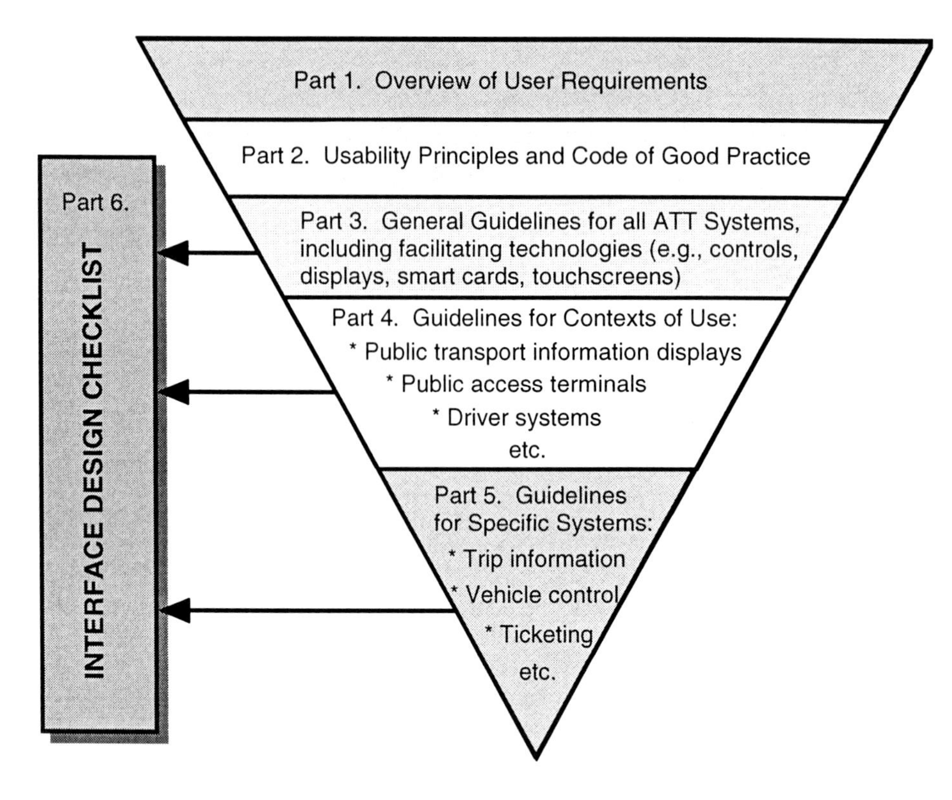


**Figure 15.2** Structure of the *TELSCAN Code of Good Practice and Handbook of Design Guidelines* (Nicolle and Burnett, editors, 1999)

Part 1. Overview of User Requirements
This section presents a summary of the user requirements from the TELSCAN project's data-capture activities.

Part 2. Usability Principles and Code of Good Practice
This section discusses the concepts of usability and user-centred design and their application to designing for older and disabled travellers, emphasising that adherence to these principles constitutes good design for all users.

Part 3. General Guidelines for all ATT Systems
This section specifies guidelines which are applicable to all systems for older and disabled travellers, whether by car or by public transport, with regard to controls, displays, smart cards, the Internet, training/documentation and other physical characteristics of the system. Many of these general design guidelines draw from the ones presented in this book, e.g., the *Nordic Guidelines* (Nordic Cooperation on Disability, 1998) described in this volume.

Part 4. Guidelines for Contexts of Use
This section identifies key design issues with respect to the context in which the telematics system is being used. This covers, for example, when the system is used

as an interactive public access terminal, non-interactive public transport display signs, in the driving context, or in multi-modes (i.e., changing the form of transport, for example driving to the station and then travelling by train).

Part 5. Guidelines for Specific Systems and System Functions
This section identifies elements of the travelling task which can and should be supported by ITS. A designer can choose the function which is performed by the system under design or evaluation. Alternatively, the designer can choose a specific system if such guidelines are available and relevant. For example, the function 'trip information' is provided to the traveller through route guidance and navigation systems. With respect to the function of 'vehicle control,' particular guidelines for adaptive cruise control systems and collision avoidance systems have also been proposed.

Part 6. The TELSCAN Interface Design Checklist
The checklist (Table 15.1) is a summary of the most important guidelines contained within the *Handbook*. If taken into consideration, it is expected that the usability of an ITS for older and disabled people would be significantly improved. References to *Handbook* sections are provided, so that the reader can discover the rationale behind specific guidelines and observe relevant examples.

Many of these guidelines are still too general and more specificity would better assist the designer to include the needs of people who are older or disabled. Furthermore, there are certain aspects of the travelling environment where it is not easy to apply general guidelines without some additional information, interpretation or testing. This is particularly true of systems used in cars, where the travelling task is complex, the primary control task is very demanding and the environment is constantly changing. For these reasons, guidelines in the travelling environment ought to be based on highly valid, very applicable, often replicated and consistent data, but this is usually not the case (Landau *et al.*, 1998).

Our work emphasises the fact that, although general guidelines are important, the most useful and specific guidelines for systems for travellers emerge only when carefully chosen research questions are investigated. The following example will illustrate this point:

In simulator testing of adaptive cruise control (ACC) systems, it was found that drivers with lower limb impairments, using hand controls for accelerating and braking, prefer a longer average headway, or distance to a leading vehicle, than that which the system used (Nicolle and Peters, 1999). The study found that the mean headway was approximately 0.7 second shorter when drivers were using the ACC system. Since an often-used headway value is 1.4 seconds, it was recommended, as a starting point, that the value should be prolonged to 2.1 seconds. This led to a specific guideline recommending that the headway for such systems be individually adjustable, by a qualified specialist, according to the driver's characteristics and preferences. However, people with different types of impairments should be included in further testing to determine what the adjustable range should be, and, in addition, testing in real traffic is needed in order to validate this guideline. We look forward to more such testing in the future.

**Table 15.1** Sample page from *Interface Design Checklist* (Nicolle and Burnett, editors, 1999)

| Displays | Tick if YES | Guidelines Handbook References |
|---|---|---|
| Alternative means of output (e.g. visual/auditory/haptic)? | | 2.1 (Usability Principles) |
| Displays conform to expectations? | | 2.6 (Usability Principles) |
| Unrestricted access to displays (e.g. flexible locations, no obstacles/restraints)? | | 3.2.2 (General)<br>4.1.1 (Kiosks)<br>4.3.4 (Driver Systems) |
| Displays easily viewable by a wheelchair user? | | 3.2.2 (General)<br>4.1.2 (Kiosks) |
| No need to peer over glasses to view/read display? | | 4.3.5 (Driver Systems) |

## 15.4 GUIDELINES FOR TRAVEL INFORMATION

All travellers need relevant information in order to carry out a journey safely and in comfort, whether by public or private transport. This information must be correct and easily available, even more so for people who are older or disabled, who often would not attempt a journey unless that knowledge is known beforehand. However, without looking closely at the requirements of each and every traveller, it is difficult for designers to know precisely what information is needed.

Travellers may need to know:

- which buses have a low floor for easier access
- whether there are reduced fares for people who are retired or disabled
- how long it will take to walk from one platform to another
- if there are any stairs to climb
- whether or not assistance is available while boarding a train
- if the more economical way of travelling from the airport to the city will be inaccessible in any way
- whether the mode of transport, stations and terminals, and all their facilities, are accessible to people using wheelchairs, and that the weight and dimension of every type of wheelchair can be accommodated

- how to book or reserve a seat and the method(s) of payment, along with alternative ways of doing so (e.g., over the telephone, by fax, or by credit or smart card).
- whether information and ticket offices have induction loops for people with hearing aids
- whether there are audible and visual announcements, as well as alternative navigation guidance for people with visual impairments
- whether there are accessible waiting rooms with visual and audible information on transport services
- whether stations and terminals have accessible parking spaces, toilets, refreshment facilities, telephones and help points, whether they are staffed and during which hours.

To illustrate, travellers with mobility-related impairments encounter specific difficulties in discovering the procedures and layout of airports, train stations and bus stations prior to making a journey. Information systems accessible from home, at travel agents or in other public places may help alleviate this problem, possibly by employing virtual reality technology to provide a 'dry run' visit of unfamiliar places.

Relevant information can also be provided to travellers through the use of 'smart' cards, which not only act as a control mechanism but can also provide information on a person's specific travelling needs. For example, bus passengers with visual impairments would particularly benefit from information which would help them locate the door of the bus. In this respect, a 'contactless card,' which can work at a distance of up to 10 cm (see Chapter 14 and www.tiresias.org/pats), could request a sound to be made (such as a repeated tone) close to the bus door. A smart card could also enable a request to be made for automatic lowering of the floor of a bus or for a ramp to be lowered. Both of these would help an older or disabled traveller to enter a bus but would also assist a parent carrying a small child, or anyone with heavy shopping.

Designers would benefit from a list suggesting the type of information which could and should be provided to meet the special needs of older and disabled travellers. The TELSCAN project has put this knowledge into *a Travel Information Checklist* (Veenbaas, 2000), which is available through the project as a working prototype (Figure 15.3). Designers of systems can use this to identify the most important travel information items that ideally ought to be included in a particular system.

Using the database of design guidelines, a designer should be able to identify the interface design guidelines for a particular system (e.g., an information system), in a particular context (e.g., a kiosk), and then identify particular information units that need to be included in the system so that older or disabled travellers have all the information they need to set out or continue on a journey. Without this sort of information beforehand, many people who are older or disabled will not even attempt to travel, even if the station *is* accessible. However, it is clear that the task of including all the required information on the transport infrastructure and all its facilities, in every type of intelligent transport system, would be a long and expensive task, and also one that would need to be updated on a regular basis. If instead inclusive design could one day become the norm, then this would seem to be the most sensible and cost-effective solution.

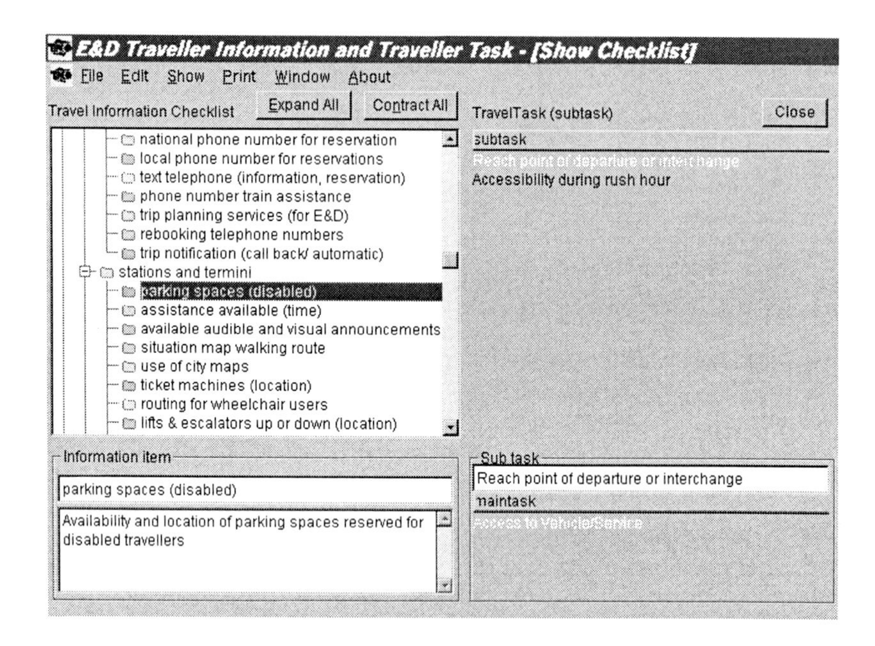


**Figure 15.3** Extract from *Travel Information Checklist* (Veenbaas, 2000)

## 15.5 GUIDELINES FOR EVALUATION

### 15.5.1 Overview of assessment methodology

Following interface design guidelines and including relevant information will not in itself guarantee a perfect system. User-testing must also be done, following a methodology which will include people with different impairments who may be likely to use the system. By considering the users' requirements, TELSCAN is also able to provide assistance to designers in this area.

As noted above, system evaluators may in principle wish to include older and disabled people in their evaluation plans but may need guidance on which groups to include and whether specific methods or testing environments are appropriate. An *Assessment Methodology* developed by the TELSCAN project (Marin-Lamellet *et al.*, 1999) makes recommendations on the human factors issues which the evaluator needs to consider. This includes advice on which user groups to include in the evaluation and also advice on testing in particular environments or contexts (see overview of the methodology in Figure 15.4).

The first aspect of the methodology gives advice on which users should be included in the usability assessment. To this end three indicators were developed, the ordering of which does not suggest that one indicator is more important than another or that the activities need to be performed in a prescribed order:

- Mobility Indicator, i.e., the extent to which system functionality could especially help people with particular impairments overcome their mobility problems. For example, a person using a wheelchair could particularly benefit from increased knowledge to aid in trip planning.
- Problem Indicator, i.e., the difficulties that might be experienced by people with different impairments when using the system. For example, a system which utilises a visual display containing text may be a problem for those with poor sight and/or language skills (see Table 15.2).
- Extent Indicator, i.e., the number of people who will potentially be affected by the system. For example, it is important to remind designers that by the year 2021, there will be a significant increase in the proportion of people over 80, and a reduction in the young adult population (Kumar, 1997).

Notice in Table 15.2 below that older people are always recommended to be included in the assessment. Even though when following a functional classification, they would not form a separate group for data capture or assessment, the requirements of a person who is elderly with mobility problems can be very different from a young person in a wheelchair. We, therefore, suggest that elderly people are included in all assessments as a matter of course.

**Table 15.2** Extract from the Problem Indicator: User groups for which particular system interface characteristics could cause problems (Marin-Lamellet *et al.*, 1999)

| **Sensory modality** | **Tick if YES** | **User groups** | |
|---|---|---|---|
| Is input required with hands/fingers? | | UL, C/D, F | + Elderly |
| Is input required with feet/lower limbs? | | LL | + Elderly |
| Is input required with speech? | | L/S | + Elderly |
| **Location of displays/controls** | **Tick if YES** | **User groups** | |
| Is access to any displays/controls restricted (e.g. fixed locations)? | | LL, UL, A | + Elderly |

**Key** to User Groups with various impairments

| | | | | | |
|---|---|---|---|---|---|
| **LL** | Lower Limb | **C/D** | Co-ordination/ Dexterity | **L/S** | Language and Speech |
| **UL** | Upper Limb | | | | |
| **UB** | Upper Body | **F** | Force | **SL** | Sudden loss of Control (visceral) |
| **A** | Anthropometrics | **V** | Vision | | |
| | | **H** | Hearing | **C** | Cognitive |

The second aspect of the assessment framework provides advice on the particular context in which the system is used and the choice of the testing environment, as some systems could be unsuitable for persons who are older or disabled. For example, when testing a system with drivers having lower limb impairments, it is necessary to consider whether the driver uses hand controls for accelerating and braking. If the driver is not using his or her own car, hand controls for accelerating and braking may need to be installed or simulated, and may need to be different, or configured differently, for different drivers. The assessor needs to investigate well in advance what car adaptations may be necessary and to discuss the type and costs of car adaptations with experts to see if purchase and installation are feasible (for example, see Nicolle and Peters, 1999).

At the next level, the framework provides guidance on choosing the methods and tools which will be most suitable for use with persons who are older or disabled, as well as the protocols and ethical issues which need to be considered. For example, focus groups with people having hearing impairments need to consider whether a sign interpreter would be helpful, both for signing and to assist in understanding speech. Also, including older or disabled drivers in simulator testing must consider various issues such as accessibility of the simulator to people with lower limb impairments, the use of car adaptations, fatigue and simulator sickness, all of which can either cut short the testing or eliminate certain users entirely (Nicolle and Peters, 1999).

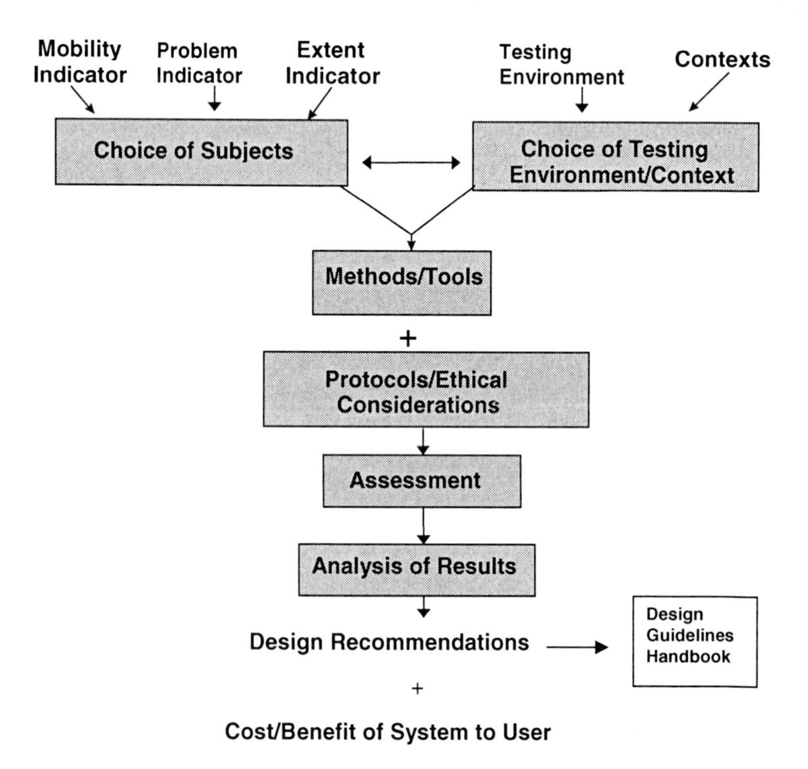


**Figure 15.4** Overview of *Assessment Methodology* (Marin-Lamellet *et al.*, 1999)

As some travellers may tire easily, the testing time may need to be limited. The assessor should consider the distance that the subject may have to travel to the test site, as this may have implications for either the choice of volunteers or the choice of test site. The assessment framework also reminds the evaluator that private and public transport vehicles, their environment, simulators, laboratories, nearby toilets and other facilities must all be easily accessible. And, of course, travellers must always know they can stop the assessment when they wish and that all results will be confidential and the user anonymous.

As well as providing advice on the procedure for analysing results, two case studies are given in order to illustrate the framework. Finally, some further guidance is given to designers to help them put the Mobility, Problem and Extent Indicators into practice by way of a brainstorming session.

### 15.5.2 The brainstorming process

It is important to emphasise that the methodology does not choose the user groups to include in the evaluation—rather it highlights what the relevant issues are when making this decision. The Mobility Indicator, the Problem Indicator and the Extent Indicator have to, therefore, be analysed and tailored according to different considerations, and this is best accomplished by way of a brainstorming session.

Ideally, no user groups should be excluded, but with resource constraints it is always necessary to make certain difficult decisions and trade-offs. This methodology does not suggest that the difficulties of certain people with disabilities are not severe enough to include them in the assessment process. It merely attempts to assist designers/assessors in making realistic and cost-effective plans to ensure that the system is more usable by more people, within their resource constraints. So we recommend that the assessment team conduct a brainstorming session to interpret the data derived from the three indicators in order to arrive at the best possible choice of user groups.

Brainstorming is a technique by which people from different backgrounds (e.g., the members of a multi-disciplinary design team) come together to inspire each other and generate new ideas (Poulson *et al.*, 1996).

The starting point of brainstorming is important and the group leader must present the issue or question to the others. We recommend that the focus of this brainstorming session is:

"*How do we interpret the data derived from the three indicators in order to arrive at the best possible choice of user groups to be involved in the system's assessment?*"

Some key issues which members of the team might consider are listed below and these could be presented by the group leader before the session. However, these issues are in no way meant to limit the opinions or creativity of the participants.

It is necessary to consider whether the transport system being assessed is accessible to all user groups and also whether it contains relevant information for those user groups. For example, even though people with lower-limb impairments may be selected through the Mobility Indicator because they could especially benefit from a trip information system, it may not be possible to include them in the assessment if the transport vehicle itself is not accessible. Using the Mobility Indicator, therefore, highlights the importance of accessibility.

It is necessary to consider the adverse effects and consequences that a system may have on one group, which might at the same time be benefiting another group. For example, mobile telephones are wonderful for people using wheelchairs but for people using hearing aids, mobile telephones may not be acceptable if they cause interference.

In choosing user groups for assessment, it is also necessary to consider the type of travelling tasks that may be problematic for persons with different levels of visual acuity. For example, in the UK a driver must be able to read a standard number plate (containing letters and figures 79.4 mm high) in good daylight from 20.5 m (67 feet), using glasses if necessary. If glasses or contact lenses are needed to do this, the driver must wear them every time he or she drives. And so, if a user group does not have the required level of visual acuity to drive, they need not be included in testing the usability of a particular vehicle control system. On the other hand, a person with other forms of visual impairment, e.g. colour blindness, may have problems if the system's display relies on colour-coding and so should participate in the user trials.

It is also important to consider user groups with multiple impairments, which might make the use of the system that much more problematic. Examples where combinations of impairments are likely to occur are:

- People with lower limb and upper limb impairments
- People with visual and hearing impairments
- People with cognitive and mobility impairments.

Therefore, if the Problem Indicator suggests that people with, for example, visual and hearing impairments ought to be included in the assessment, consider trying to include users with not only these individual impairments, but also a combination of these.

When the safety of travellers is an issue (e.g., for people who have visual difficulties but still meet the requirement to drive) the Problem Indicator becomes even more important. It would then be necessary to investigate the displays and controls that might cause particular problems for that user group. Issues identified through the Problem Indicator will often help to identify not only the experimental design and scenarios for testing, but also the need to consider redesigning the system's interface, using the design guidelines developed by TELSCAN.

Discussing issues such as these will help to choose the most relevant groups to include in the assessment strategy. It is necessary to stress that the team must make the final decision and these are only tools to highlight important issues. Furthermore, in order to implement these decisions, it is necessary for assessors to know how to find people who have different types of impairments. More work is needed to extend this methodology to include a list of contacts to help designers find the right organisations and individuals, who will often be only too glad to assist.

## 15.6 SUMMARY AND THE WAY FORWARD

This chapter has provided an overview of design guidelines for intelligent transport systems for travellers:

- Prescriptive guidelines on how to design the interface between the user and the system
- Prescriptive guidelines on what type of specific information is needed by older or disabled travellers
- Process guidelines including methods and tools to evaluate systems for ease-of-use by older people or people who have disabilities.

The design of systems for travellers should, of course, follow the same general human factors principles which apply to other systems, for example for accessible computers in an office, in a home or in public places. Hence, the general guidelines proposed throughout this book are also relevant to many systems for travellers. A kiosk designed to provide travel information in a train station needs to be accessible to a person in a wheelchair or with a visual impairment. Likewise, information on train times and services on the World Wide Web needs to be accessible by providing a text-only version. It is also important that Web sites or information kiosks provide relevant information for travellers who are older or disabled, or possibly that the information can be provided in a portable fashion to be accessible whilst the person is on the move. Such recommendations are part of the added-value of the work done in projects like TELSCAN.

However, even when designers have guidelines in an accessible format (paper- and/or Web-based), there is still much work to be done. Further increased awareness is needed so that designers know not only which guidelines are available but also when to use them. The ITS industry is often not aware of the guidelines emerging from the assistive and rehabilitation technology sector (e.g., in Chapters 9, 14 and 16). To facilitate this process, developers of guidelines should work together in co-ordinating their efforts and integrating their guidelines (e.g., through linked Web sites), thus providing a 'one-stop shop' for designers wherever possible.

## ACKNOWLEDGEMENTS

The authors wish to thank the Commission of the European Communities for partially funding the work of the TELSCAN project, and the entire TELSCAN consortium for their contributions to this study:
Aristotle University of Thessaloniki (GR), TRD International (GR), National Technical University of Athens (GR), Mobility International (BE), HUSAT Research Institute, Loughborough University (UK), TNO/Vehicle Dynamics Department (NL), TNO/Institute for Perception (NL), De Langstraat (NL), University of Stuttgart/IAT (DE), Swedish Road and Transport Research Institute (SE), AMU Gruppen (SE), Lund University (SE), INRETS-Lesco (FR), Technical University of Lisbon (PT), Cranfield University (UK), and Traffic Solutions Ltd. (UK).

CHAPTER SIXTEEN

# Guidelines for the Development of Home Automation Products

David Poulson

## 16.1 INTRODUCTION

The guidelines collated here are drawn from a variety of sources,[1] covering issues of basic physical accessibility to the home environment as well as the design of interfaces to home automation systems and other products. Home automation is still an emerging technology, however, and relatively little is known about how to design and evaluate these technologies effectively. Few products are in the marketplace and a lack of standards in this area has also created significant problems in the actual construction and operation of such systems. Sophisticated home automation systems are still very much a rich man's toy or an enthusiast's plaything and in addition most developments are experimental in nature rather than being well-established technologies. A number of experimental systems have been developed but evaluation as to their long-term value has been limited.

Gann *et al.* (1999) also raised these issues in their description of two pilot developments that took place in the UK recently, funded by the Edinvar Housing Association and the Joseph Rowntree Foundation. The authors concluded that a number of factors were limiting the take up of home automation technology, including:

- technology being immature
- a lack of standardisation
- a lack of cheap products
- a poor understanding of user needs.

However, it is misleading to think of home automation as a completely new field of investigation, as most homes have some degree of automation to control home functions, e.g., central heating systems. Home security systems are also relatively common, taking information from a number of intruder sensors mounted in the home and integrating them into a common monitoring and alarm system.

In the past there has been little integration of the different systems in the home and this is where the vision of home automation in the future differs from existing home products. Such integration is a challenge for design, as it is the

---

[1] This chapter largely draws on work carried out as part of the European Union-funded CASA project (Concept of Automation and Services for people with speciAl needs—TIDE Project 1068) and other sources.

performance of systems made up of a large number of elements which is of interest and the performance of the whole requires careful integration of each part. For example, a home security system might consist of a number of passive infra-red sensors connected to a common network. Each sensor must work in isolation for the security system to operate effectively but in addition has to form part of a wider network providing effective coverage of the home. For the security system to work effectively, the optimal positioning of sensors in relation to other sensors has to be selected and this will be determined by the layout of the particular home where installation will take place. The presence of domestic animals can also influence this decision, along with the types of sensor to be used.

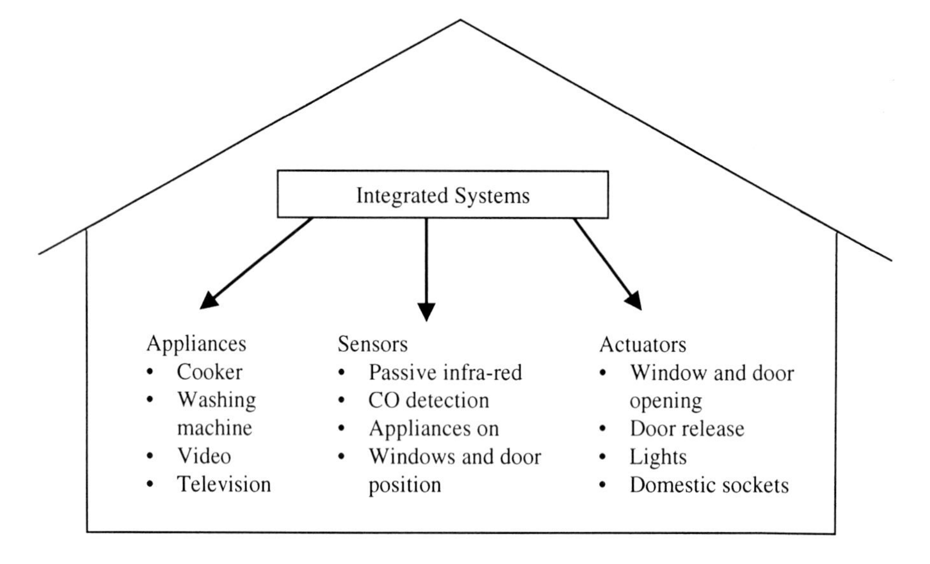


**Figure 16.1** Integrated systems in the home

In an integrated home automation system (Figure 16.1) the situation can be even more complex, as individual elements of the home automation system can be used for a variety of purposes and may form part of a number of functional sub-systems. For example, the passive infra-red sensors which were used as part of a security system may also be used to detect any activity in the home and as a way of locating where the occupant is in the home.

In addition, a larger number of devices in the home can be linked to form an integrated system. These can be household appliances, sensors mounted in the home and actuators. Currently household appliances are largely isolated devices, but increasingly there is talk of ensuring that domestic appliances can communicate as part of home networks. For example, the emerging Bluetooth protocol[2] is being promoted as a relatively simple way of facilitating wireless communication between household appliances and allowing them to be connected easily to create integrated systems.

[2] See bluetooth.com for more details.

## 16.2 THE VALUE OF GUIDELINES

Guidelines are by their very nature simplifications; they are advice drawn from best practice, which in the majority of cases will provide satisfactory advice for making design decisions. However, the weakness of all guidelines is that they are generalisations based on best available knowledge and draw on the collective experiences of those putting together such guidelines. Where technological innovation is taking place, such guidelines do not exist, and drawing on best available knowledge from existing technologies may be both limiting and inappropriate. For example, current guidelines for visual displays are largely based on the cathode ray tube as the display technology and the limitations that this imposes in terms of size and positioning of the display. With cathode ray technology there is always a bulky and heavy tube that uses high voltages and generates significant heat. The space taken up by the tube makes it almost impossible to mount close to the horizontal and its weight also means that it has to be in a relatively fixed and rigid position. Current guidelines for workstation design are based on these limitations, which as technology advances will become outdated. Flat-screen technology is now relatively common for portable computers and is likely to become more common for a range of other applications too.

Another limitation of guidelines is that they can also be influenced by culture to a greater or lesser extent and may embody assumptions as to how activities should be carried out. For example, it is normal in the UK for light switches to be mounted higher than door handles, whilst in other countries it is normal for them to be at the same level. Toggle light switches in the USA are off when put in the down position, whilst in the UK this indicates that the light is on. Such variations make it difficult to develop specific recommendations that have broad applicability and often it is better to consider national standards and guidelines rather than attempting to create standards and guidelines with universal application.

Guidelines also need to be considered in the particular cultural and social setting in which they will be applied. For example, the design of assistive technology products in the UK has been dominated by a medical rather than a consumer-oriented perspective, largely due to extensive state funding of medical and social support. This has led to medical safety being one of the overriding design principles for assistive technology products, with consumer opinion and attitude being seen as of secondary interest. This is now changing as the UK moves towards a more market-driven approach to medical and social support, but it still influences how design in this area is perceived.

It is therefore important to look carefully at the assumptions that have acted as the basis for any guidelines and to critically assess whether these apply in the specific design context being considered. Guidelines may have been developed with a very specific user group, activity and culture in mind, and may therefore have been generalised beyond the limit of their validity. The context in which a guideline was developed may have been lost by its transfer into another application domain and the validity of the guideline may often not be questioned in this process. Guidelines may also have been developed from a very limited body of knowledge which may not have been made clear by its original author, or may have become lost over time as other authors cite the work.

For example, a number of authors have provided recommendations regarding the maximum amount of force that should be used to operate a simple switch to make it accessible to people who are older and disabled (e.g., Scadden and Vanderheiden, 1988; Nordic Cooperation on Disability, 1998). Many of these recommendations are similar, but what is less clear is whether they represent independent research or are drawn from a common source. A number of authors providing common recommendations may give a spurious impression that, as a result, the guidelines must be more valid. The reality may be more a reflection of the limited amount of research into the topic concerned and the fact that the authors have all drawn from a common source of information.

As home automation is an emerging technology, it is impossible to find a coherent body of knowledge and guidelines specific to this sector; rather advice is based on limited numbers of examples and extrapolation from related areas such as the design of consumer and IT products. As remarked earlier, home automation is also marked by the degree of integration needed, where individual devices inter-operate and together make up a working system. Such complex systems do not lend themselves to design by simple guidelines and the developer must always question the validity of such generalisations in the context of the particular application being developed.

## 16.3 DEVELOPING PRODUCTS AND SERVICES IN HOME AUTOMATION

The range of products and services that can be developed for the home environment is almost infinite, and different products and services will need to be closely tailored to individual needs. For example, a frail elderly person may require home systems which monitor the status of the home and alert outside agents to problems in the home, whilst for other user groups different services may be appropriate such as home control applications. One of the primary distinctions in service-need relates to whether physical or cognitive support is needed, but again this is a simplification as needs are also directly related to the aspirations of the user, and to the quality of supporting infrastructures and geographic location. A physically disabled person is likely to need a very different kind of home automation system than a person suffering from mild dementia. In the former case, emphasis is likely to be on the need for control applications, whilst with the latter, home safety monitoring is likely to be emphasised. Whilst it is impossible to create a definitive list of services which are needed for different design contexts, Table 16.1 gives examples of some of the applications of this technology that have been developed, or are being explored in pilot applications.

## 16.4 SOME HIGH LEVEL DESIGN OBJECTIVES

Few guidelines for the development of home automation products for use with older or disabled people exist and most literature is of a more general nature covering the design of assistive technology products rather than specific home automation applications. The following high level design objectives are drawn from

**Table 16.1** Applications of home automation

| Area of Need | Automated Home Solution |
| --- | --- |
| Accidents | Active alarm system operated by alarm pendants or buttons installed in the home |
| | Passive alarm system detecting inactivity of clients in the home, e.g. lack of movement, use of water, fridge, etc. |
| | Remote monitoring of health status—blood pressure, body temperature, heart rate, etc. |
| | Remote video telephone or CCTV monitoring of home |
| | Automatic lighting to detect movement |
| Home security | Intruder alarm systems linked to remote service centres |
| | CCTV monitoring of entrance to property and door bell use |
| | External lighting activated by movement detection |
| | Doors and windows to property being open |
| | Intruder systems automatically enabled by leaving the property |
| Home safety | Water leak, gas leak, CO and smoke detection. Detection of critical appliances being left on, e.g. cooker |
| | Alarms for fridge and freezer defrosting |
| | Isolation switches to turn off all power to the home. Other utilities such as gas and water may be controlled in a similar fashion |
| | Automatic switch-off of utilities in the event of an emergency, e.g., water leaks, gas leaks |
| | Status monitor for home, showing condition of critical appliances |
| | Intelligent appliances reporting need for maintenance and repair |
| Home safety/economy | Timers to automatically switch off appliances after a period of operation |
| Home economy | Energy-management systems for the use of electricity, i.e. at off-peak times |
| | Energy-management systems for heating the home—timers and complex sensors |
| | Detection and display of abnormal or unusual conditions in the home, e.g. door bell ringing, telephone ringing, potentially dangerous appliances left on, etc. |

| Area of Need | Automated Home Solution |
|---|---|
| Control of home | Intelligent sockets for all domestic appliances (switching and status of power usage) |
| | Control of switched appliances from a central control console |
| | Remote control of television, video, hi-fi and radio |
| | Remote control of lighting |
| | Interfaces to specialist control units and communication devices, e.g. speech recognition units, scanning interfaces, etc. |
| | Hands free intercoms and telephones |
| | Remote door release and openers |
| | Remote window openers |
| | Electrically adjustable beds and chairs |
| | Remote control curtains/blinds opener |
| | Remote control of garage doors |
| | Control of home from a remote location (service centre, office, etc.) |
| Support of client | Tele-alarm service extended to cover advice and other support |
| | Reminders for appointments and medication |
| | Video-telephone link for advice and emotional support |

a variety of sources including Batavia and Hammer (1990), Sandhu (1993), Poulson and Richardson (1994), and Poulson (1997).

### 16.4.1 Suitability for task

Products need to be carefully assessed in terms of whether they provide demonstrable benefits to their users compared to alternative ways of performing the same tasks.

Developments in home automation have historically been driven by the vision of a push-button home, where household activities are achieved by the single push of a button with the support of complex actuators, e.g. door and window openers. However, there is little evidence that such degrees of home automation are necessary or desirable for most people and unless a person suffers from a severe physical impairment the value of many remote-control applications is questionable. For example, it is technically feasible to develop remote-control systems for cookers and washing machines but harder to see how such controls might be of value to the consumer.

Home automation technology is also currently limited in terms of the household tasks to which it can be applied and often it is the physical tasks that older and disabled people find difficult which cannot easily be supported by its use.

For example, older and disabled people often face problems with physical household tasks such as washing clothes and cleaning the home. Whilst these tasks are supported to some extent by existing home technology such as washing machines and vacuum cleaners, it is difficult to see how home automation technology can be applied to improve significantly many of these devices without a major advance in robotics and artificial intelligence.

Work within the CASA project (Poulson, 1997) has demonstrated that there is considerable potential for using home automation technology to monitor the safety of the home and that such systems are likely to have a wider potential market than control applications. Home security and safety are important factors in older and disabled people being able to live independently and it is also one of the areas where technology can be applied relatively cheaply. However, the cost of the technology is a critical limiting factor in its likely take-up in this sector, as many older and disabled people are on low incomes, and state funding for technical aids is often limited.

### 16.4.2 Ease of use

Ease of use is a critical factor in the use of home automation products, and elderly people in particular can be expected to have a fear of any product that is perceived to be too complex. It is important to realise that a level of complexity taken for granted by many younger people can be unacceptable to many older users who have little experience of using technology. As an example, within the CASA project the decision was made to develop a home security and monitoring system that was activated by a simple lock and key (as shown in Figure 16.2), rather than implementing a keypad-based system with a four-digit security code.

**Figure 16.2** Lock and key activation of home security and monitoring system

Ensuring that the operation of products conforms to users' expectations can facilitate ease of use. Where possible familiar metaphors should also be used, e.g. using television or telephone, rather than introducing large numbers of new concepts for people to learn. This is particularly important if frail and confused elderly people are seen as a potential user group. Within the CASA project the decision was made to provide an interface to the home automation system which used a standard television and remote control unit (Figure 16.3).

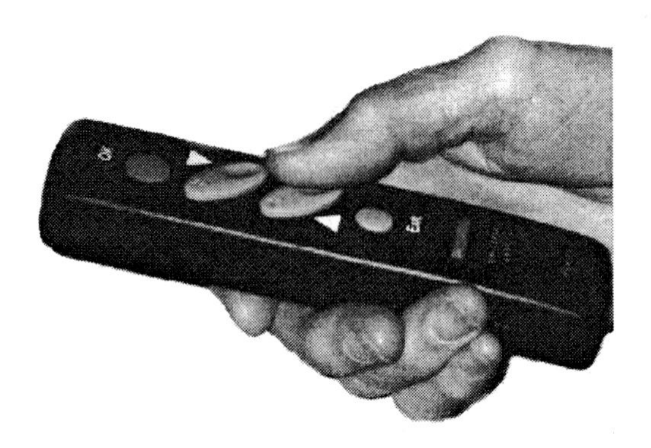

**Figure 16.3** CASA Project's remote control

Figure 16.4 overleaf shows the top-level menu screen for the Spanish version of the CASA system. The functions illustrated are:

- messaging service
- assistance or alarm service
- security system
- monitoring the status of the home
- controlling the home.

The use of a simple remote control and menu-style interaction was a design compromise used to allow flexible configuration of the product to an individual client's needs. Where there are a small number of functions, there is some indication that it is better to provide a simple interface using a number of dedicated keys compared to a more complex menu structure.

An attempt should also be made to minimise that which the elderly person has to remember in order to use a system. Recognition is a much easier activity than recall and so, where possible, the interface should give the user guidance as to what actions are permissible at any given point in the interaction rather than relying on the user's memory.

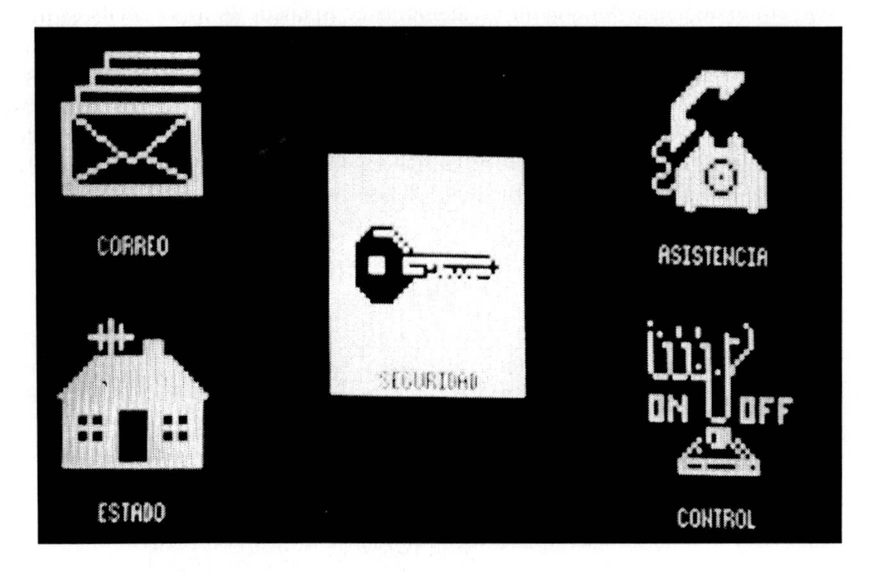


**Figure 16.4** Top-level menu screen

### 16.4.3 Simple functionality

A related trade-off that needs to be explored is between providing more functions with the corresponding possibility of a complex control structure and more simple systems with reduced functionality. The indications are that the latter strategy is likely to be more effective with older people and, where feasible, systems should be developed which only provide simple functionality.

Where possible, interfaces to home automation products should not force older and disabled users to have to respond quickly, as the person may be much slower cognitively and physically than their younger counterparts, and for certain disability groups quick and controlled motor activities are also difficult to achieve.

### 16.4.4 Providing good feedback to users

In order to improve the chances of an older or disabled person being able to detect system messages, it can also be useful to give the same message through different senses. On a fire alarm, for example, a flashing light could be used to supplement an alarm siren. Where possible, devices should give visual, auditory and even tactile feedback of their status, as the more senses that are engaged, the greater the chances that signal detection will take place.

Feedback regarding the consequences of user actions also needs to be given. In some systems this may be immediately apparent by the nature of the control task itself, e.g. the telephone being rung, whilst in other cases more explicit feedback will need to be designed into any interface. Where a control action can be delayed, e.g. commanding a window in the home to be opened, the user needs to be given immediate feedback that the task has started.

Home automation equipment should also provide some feedback to the user that it has been switched on and is working. Often this can be obvious from the way in which the equipment operates, but in some cases it can be useful to have some form of display, e.g. an LED showing that the equipment has been powered-up and is connected to the mains power supply.

### 16.4.5 Safety in use

Where possible, technology should not introduce higher levels of risk to the older and disabled person than if that technology were not present. However, product safety usually is achieved at a price and a trade-off needs to be made between acceptable levels of risk and cost. For example, conformance to medical safety standards will improve the potential safety of products in a range of operating environments but may mean that subsequent manufacture is considerably more expensive. Conformance to some medical standards of safety may also not be necessary for home use, where other products in the home environment do not conform to such standards.

### 16.4.6 Reliability and fail-safe operation

Home automation products and services need to be highly reliable and also to have fail-safe operation. Failure of home automation technology should not put users into a situation of greater danger than if they did not have the device in the first place. The user should not be placed in a position of danger due to failure of equipment or failure in the environment, e.g. power cuts. In the design of environmental control systems which operate critical home operations (e.g., use of telephone, raising alarms, opening door to property), it is important that these activities can still be performed in the event of a system failure. Thus it can be important still to have manual ways of performing the same tasks and/or to have back-up systems to cope with emergencies. For example, it is common practice for developers of environmental control systems to provide battery back-up of critical functions so that parts of the system are still operational in the event of a power failure to the home.

Other examples include ensuring that electrically opened windows can still be operated by hand and that electrical sockets as part of a home automation system are also fitted with a manual override.

### 16.4.7 Discouraging passivity

An image often associated with home automation is that of the push-button home where everything is done for the user. However, there is a danger of providing too much support for older and disabled people in their homes and a careful assessment of what they need, rather than necessarily want, is required. Generally home automation products should not encourage passivity and in addition should not foster over-reliance on their use. For this reason, it can often be better for systems to remind users to perform actions, rather than performing those actions directly on the users' behalf.

### 16.4.8 Design aesthetics

The styling of home products is anticipated to be critical in determining their acceptability to older and disabled people and this is often a neglected area of product development. It is recommended that products be professionally styled to provide an attractive and acceptable image. For elderly people, it is probably better to provide classically designed products which avoid the use of overly bright colours. The converse is likely to be true when developing products for use by the young disabled person and a clear segmentation of market sectors needs to be identified in order to ensure that products are developed to take different user groups into account. Providing a choice of product styling and colour should also be considered.

## 16.5 SPECIFIC DESIGN GUIDELINES

There are a variety of sources of information covering the design of the home and its products and it is clear that these have relevance for the design of home automation systems for use with older or disabled people. Commonly, such recommendations are based on the principle that access to controls and displays should be possible for wheelchair users, as such users have a lower reach envelope than a standing person. It is also possible to list more specific design criteria for developing interfaces to home automation products. These design guidelines for controls and displays draw from those general HCI guidelines as discussed elsewhere in this volume.

### 16.5.1 Architectural and accessibility features

Goldsmith (1984) is still seen as one of the definitive sources of design material in this area, covering recommendations regarding the architecture of buildings for disabled people, as well as more specific recommendations on fixtures and fittings. Another very useful source of design information regarding the design of housing interiors for older and disabled people is given in Raschko (1991).

Whilst it is not possible to review the architectural features of designing property for older and disabled people, some of the accessibility guidelines for electrical fittings are worth mentioning. They are relevant to the development of home automation systems that are electrically operated and may involve some changes to the infrastructure of the property, e.g. rewiring and provision of power sockets. Light switches, circuit-breakers, heating controls and entry telephones should generally be at a height of approximately 1 m above floor level. Goldsmith (1984) recommends a maximum height for a switch to be 1.143 m from floor level.

Rocker action switches (Figure 16.5) are also recommended rather than conventional toggle switches. Self-illuminated switches should also be provided at bedroom doors, bathroom doors, entry doors and any areas that are usually dark.

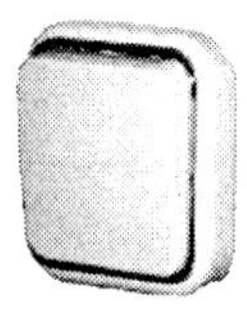

**Figure 16.5** Rocker action switch

Where possible, power sockets should be at the same height as light switches and door handles, to allow access by wheelchair users and also to assist elderly people in reaching them. Goldsmith (1984) argues that they should not be lower than 0.5 m from the ground.

### 16.5.2 Recommendations for the design of controls

Interfaces to home automation products may be through dedicated control units. However, they may also use more generic technology such as a personal computer, where that is used as the heart of the home control and monitoring system. In such cases the mouse input device may cause particular problems for many older and disabled people due to the fine motor control required. Some people with impaired hand functions will find using a mouse difficult and, for this reason, it is recommended that keyboard alternatives to the mouse, i.e., cursor keys on a keyboard, also be provided. In addition, the mouse should be operable by right- or left-handed individuals. Another option that can be explored for such interfaces is the use of a touchscreen rather than a conventional keyboard and mouse. This has the advantage of allowing a much more 'natural' selection of menu items, by merely pointing at those items required on screen rather than having to engage in a data-entry activity. In the past, problems with screen calibration were common, but now this technology is much more reliable. However, it still represents an expensive design option for home use and so is not really viable on economic grounds.

Speech-recognition technology has advanced considerably in the past few years and may also be worth considering where other forms of input are difficult. Large-vocabulary, continuous-speech systems are now available at a reasonable cost for the domestic market, but they still suffer from problems that limit their use within the home environment. High accuracy is only achieved for systems trained to the voice of the individual user and environmental noise can interfere with recognition accuracy and performance. Whilst this technology is continually being refined and developed, currently such systems are better used in quiet environments such as offices, and where a microphone can be mounted close to the mouth and at a fixed distance from the speaker.

Interfaces to home automation products are likely to be a combination of local controls mounted on or near equipment and more central units which are part of the integrated home automation system. These in turn may be operated by fixed controls, or be operated by remote control units.

In all cases, operating keys for systems to be used by elderly people need to be relatively large, with one study showing buttons of about 3 cm in diameter to be optimal (Mann *et al.,* 1994). This clearly reduces the possible number of keys for a hand-held control unit and opinion appears divided on the minimum key size that should be adopted.

### 16.5.3 Layouts of controls

In addition to the purely physical characteristics of isolated controls, it is important that controls be easily distinguished from the equipment's casing. For this reason it is recommended that high levels of contrast between controls and the background be used.

Many aspects of control and display design are likely to be specific to a particular application but there are some general principles of human perception which are relevant to consider when designing control and display layouts. More detailed recommendations for the design of interfaces are given in Bailey (1982).

- One of these is the principle of proximity, which suggests that elements of a display/control panel positioned near to each other will be seen as belonging together. Thus a cluster of four buttons on a control panel will be perceived to be associated with each other, whereas if they are spaced apart this will not be the case.
- Controls should be grouped on a control panel according to use, with critical and frequently used controls being within easy reach. Controls that are used infrequently do not need to be positioned within easy reach and may interfere with normal usage of the equipment if they are placed in a prominent position. (That is why the on/off switch and channel selector is usually displayed on the front of most television sets, while the other controls are often hidden behind a panel.)
- Controls can be grouped according to their function, i.e., controls for related activities should be positioned close to each other. This can be offset against an additional requirement to attempt to group controls together according to sequences of use, and the designer needs to experiment with different control layouts in order to reach a suitable design compromise. Both of these options should be explored but the former is likely to be more important for remote control applications.
- Another principle is that of similarity, which suggests that elements that are uniform in shape and colour will be perceived as being associated with each other. This can be used to good effect in the design of interfaces, where the use of colour and different shaped keys can make it clear that keys serve different functions. Keys can be grouped by the careful use of colour on the keys or with any labels associated with them. It is important, however, that defects in colour

vision are accounted for if colour is used and also that certain combinations of colour are avoided. Red text on a green background, for example, is a particularly bad combination and red text on a blue background is also poor.

### 16.5.4 Recommendations for the design of displays

A key recommendation for the design of displays for use with people with disabilities is that, where possible, there should be redundancy in the display medium used. Thus people with visual impairments require information to be presented via tactile or auditory cues and, conversely, people with hearing impairments are more likely to rely on visual cues. It is also important that warnings and similar alerting messages remain on the display for a sufficiently long time to be discovered by the user and one suggestion is that visual messages remain until actively acknowledged.

Because of the high incidence of visual problems in elderly people, it is recommended to use large characters on visual displays. However, the absolute size of text needed is dependent on viewing distance from the control panel and display and so no absolute recommendations for size of text and display can be given. For a viewing distance of 600 mm (normal computer usage), it is recommended that characters be approximately 6 mm high for use with elderly and visually impaired people. For input devices and displays that will be operated closer to the eye, smaller characters are permissible. However, it is recommended that, wherever possible, the designer err on the side of caution and make characters larger rather than smaller. Another suggestion is for users to be able to magnify part of a display image in order for them to continue using a visual display. A zoom feature allowing up to 16 times magnification is recommended for computer displays for use with people with visual impairments.

Hearing loss can cover a range of frequencies, though high-frequency loss is the most common effect of ageing. For this reason, auditory displays need to be adjustable in terms of volume and, if possible, frequency. Because high-frequency loss is more common, it is recommended that if beeping tones are used they have a strong component between 300 and 750 Hz, as these are easier to hear for many hearing-impaired people. Induction loop coupling should also be considered to provide improved access to auditory information and is particularly relevant where speech and music are an integral part of a systems output.

### 16.5.5 Dialogue design issues

There are no simple answers as to the most appropriate dialogue designs that should be adopted for older and disabled people to use, as the overall usability of an interface will depend largely on how the different elements of the interface interact with each other. However, for use with elderly people and those with some form of disability, some styles of interface for home automation systems are clearly preferable to others.

### *16.5.5.1 Dedicated control keys*

There are arguments that the most effective dialogue structure for the design of a control system for use with people who are elderly and disabled is to have a dedicated key for each function. This is the approach used in many existing remote control units for consumer products and has the advantage that it is relatively easy to learn to associate specific keys with specific functions. However, this approach limits the number of keys that can be incorporated into a handheld unit without making the keys too small to operate, and so, effectively, limits the number of functions that can be implemented. A study by Mann *et al.* (1994) showed that elderly people had a preference for remote controls with fewer features and larger buttons. Unfortunately, it is not clear from the report why this was the case. It is also not clear how the two designs differed and so it is not possible to interpret this result as favouring dedicated keys versus menu-driven systems. However, dedicated keys do provide some clear advantages over other types of interfaces, as users can learn to use the system with tactile and spatial cues to assist them, i.e., the position of buttons. This can be particularly important for people with visual impairments and those with learning difficulties.

One solution to the problem of there not being enough space for dedicated keys is to have shifted functions for some key operations, but unfortunately this also adds to the complexity of the interface for the user. Where shifted functions are used, it is recommended that single-handed operation is possible and that simultaneous keypresses are not needed. Dedicated keys can also be used to provide predefined 'power' functions, i.e. where a key is programmed to perform a sequence of actions rather than one, e.g. dialling a stored telephone number.

Dedicated keys do have one clear advantage over other dialogue styles in that there is less reliance on visual cues, e.g., a display is not essential to operate the device. This could be particularly useful for people with visual impairments, unless auditory displays are also included. Dedicated keys can also be made in different shapes to assist in tactile discrimination.

### *16.5.5.2 Menu selection*

Another option is to use menu-selection-style interfaces, where the user selects between options. This has the advantage of allowing the number of operating keys to be reduced and a simple control panel with four buttons relating to the functions (forwards, backwards, select and cancel) can provide all the control a menu system would need. Such a limited number of keys can make some operations difficult, however, e.g. entering numbers or letters, and for this reason this style of interaction is often used with a dedicated number pad.

Menu selection systems are generally more complex to use than dedicated function keys but are more flexible in that it is relatively easy to add functionality as a system is developed. Menu-driven systems can be constructed in a variety of different ways and decisions need to be made between having a larger number of items on each menu and having a larger number of menus with fewer items to select at each level.

Whilst generalisations are difficult, the former is probably a better strategy to adopt for use with older people, as such systems are likely to be easier for users to understand and control. It is difficult to generalise as to the optimum number of items that should appear on a single menu, as this is largely dependent on the application and how material in particular menus are organised. However, it should be remembered that menu-operated systems are inherently more complex for people with visual impairments to learn to use, unless verbal prompting is given by the system. In addition, menu systems may impose a memory load on end-users which can be inappropriate if that person has some cognitive disability. Other recommendations for using menus include:

- The use of very simple menu systems, which allow the user to build up an internal representation of the application, should be encouraged
- Simplified sequences of control activities are recommended, with infrequent or one-time tasks being placed on separate menus to the main tasks. Automatic or default settings should be provided for non-essential details
- One key access for menus is recommended, where possible
- Numerical menus should begin with 1 rather than zero, as people normally do not count from zero.

Menu selection systems can also be used with form-filling dialogue styles, i.e. where each screen has a number of fields which have to be modified and then selected. For example, the control of a domestic appliance such as a washing machine might involve selecting the type of wash required, when the machine should start, etc. It is common to fill such fields with default settings, which can then be changed by the user if necessary. Form-filling styles may be usable with elderly people, but it may be common for them to forget to actively accept default options. For this reason, form-filling styles are generally not to be recommended for use with this group of users.

#### *16.5.5.3 Other styles of interfaces*

Other interface styles which can be used are questions and command language interactions. The former are simple to operate but are long winded. This can be irritating for all but the very naive user. However, they can be useful for certain types of control systems, e.g. where speech input is being used. Due to their complexity, command language systems are generally not considered appropriate for use with most elderly people. For example, Cress and Goltz (1989) recommended presenting choices as prompts and menus to users, rather than interfaces requiring typed commands or memorised sequences.

## 16.6 DESIGNING FOR PEOPLE WITH SEVERE DISABILITY

The design of systems for individuals with severe physical and cognitive disability is largely outside the scope of this chapter but some points should be mentioned. Very simple menu-driven systems can often be the only type of interface the severely disabled person can operate, as the scanning style of interfaces can be used where the user has the opportunity to select one of a number of menu options which are successively highlighted for selection. One approach is for menus to be presented which automatically cycle through possible items, allowing the user to select a highlighted item with a single keypress. Thus it becomes possible to produce a menu system which can be operated by a single input key, e.g. a suck/blow switch. The ability to modify an interface to allow this kind of input can also be valuable in ensuring that even those with a severe physical disability can use a system. It should be noted that this interaction style is more time-consuming to operate and so is unlikely to be the most effective style for those without such severe disabilities. However, where there are few menu choices, such a style of interaction could be considered for more general use, as a single button remote control is then possible to implement. Cress and Goltz (1989) provided a number of recommendations for the control aspects of interfaces for people with severe disabilities. These include:

- Techniques which reduce the cognitive load on users should be used. These include using touch screens, light pens and touch pads as input devices, and designers should try to minimise the need for visual tracking of target items.
- Single switch activation is simplest if used for performing the same action consistently, e.g. signalling yes or no, rather than for multiple actions.
- Tasks should have motivational features built into them where possible. Incorporating game features and the use of graphics and animation can make a system much more attractive for the young physically impaired person.
- The medium of information presentation should be matched to the skills of particular users, i.e. the use of words, pictures, symbols, noise, voice or animation where appropriate.
- Very simple displays should be used, presenting few concepts or ideas on each screen and using blank space to focus attention.
- Voice output is suggested as a useful medium to use for task instructions and feedback.
- The use of transitory signals, such as auditory beeps, should be minimised. It is also recommended that these can be turned off as required.

## 16.7 DISCUSSION AND CONCLUSIONS

This chapter has provided an introduction to the range of products and services that are currently being explored in the home automation sector with particular reference to supporting the needs of older and disabled users. Many of the guidelines referred to are of a general nature, covering broad high level objectives for interface design, as well as some of the issues relating to basic accessibility. Where feasible, concrete examples taken from the work of the CASA project have also been included. However, home automation technology is still in its infancy and there is a shortage of products and services available in the marketplace. Most developments in this area have been experimental in nature and it is true to say that there is no established body of knowledge of how best to design and develop such systems. Instead, guidelines and standards primarily relate to the design of isolated components of such systems.

The range of possible products and services which can be installed in a home is almost infinite, as existing combinations of sensors, actuators and appliances can be combined in a variety of ways. True integration of the disparate technologies that make up the home is for the future and current applications in this area are to a large extent limited by the lack of available products on the market. However, it is also true to say that much work is needed in defining how integrated smart home systems should operate and the rules defining that interactivity.

Creating home monitoring and control systems which gather information on the status of the home environment allows sophisticated monitoring systems to be developed which only act on abnormal conditions in the home. For example, sensors detecting that a cooker is left on or that a fridge door remains open may be combined with other sensors which detect where the occupant is located and will only raise alarm conditions if the occupant is not in the kitchen for any length of time. To provide such a service, therefore, requires the integration of at least two sensors, as well as timers and display units in the home. As another example, a sensor detecting that a telephone or doorbell was ringing could be used to provide a television-based message to the client and also mute the sound of the equipment so that it can be heard. This would require the integration of home sensors with control of the television by the home automation system.

As more experience is obtained in developing automated home systems, we will gain a better understanding of the range of products and services that different user groups will require of such systems, as well as guidelines for integrating the home appliances, sensors and actuators which make up a home automation system.

CHAPTER SEVENTEEN

# User Friendly Software for Computer-Based Instruction and Learning Materials

Andrew Downie and Ann MacCann

## 17.1 ADVANTAGES OF COMPUTER USE FOR PEOPLE WITH DISABILITIES

Many people who have a major physical disability find that accessing learning materials via computer is essential for their independence. This is shown by several occasional papers available from the Open Training Education Network–Distance Education (OTEN–DE) which detail the adaptive technologies available for people with disabilities (Downie, 1993a, 1993b, 1995 and 1996). For instance, people who are severely dyslexic may manage well with a synthetic speech output program. Some people with learning difficulties find that reading material on screen helps them to focus on what they are reading. Additionally, people who require large print are able to work on screen after changing the font style and size to one they can use. People with visual impairment benefit from learning materials being available on computer, as described below. Many people find multi-sensory information extremely motivating. For a literature review on computer-assisted instruction for students with mild disabilities, see Anon. (1995). Okolo *et al.* (1993) provides a retrospective view of computer-based instruction, or CBI.

This chapter applies general design guidelines to the specific context of computer-based instruction and learning materials to ensure that people with disabilities can obtain maximum benefit. Special emphasis will be given to the specific features of educational materials available electronically.

## 17.2 PROBLEMS WHICH GRAPHIC USER INTERFACES (GUIs) CAN CAUSE FOR PEOPLE WITH DISABILITIES

Similar to other environments, access to educational software can be difficult for people with disabilities. Heavy use of visual and other multimedia presentations in modern educational software can cause problems for those with severe vision impairment. The three major options for these people are large print, electronic Braille displays and synthetic speech output. Purpose-designed print/graphics enlargement software allows, among other features, enlargement of information on the screen and selection of colours independent of the host software. For speech and Braille, the presentation of information from the computer screen in a sensible form requires sophisticated accompanying software, known as ‘screen readers’.

The development of these packages lags behind the development of the software which they read.

As Windows was gaining in popularity, people who are blind became increasingly concerned that they would not have access to state-of-the-art computing. However, software producers did provide some quite effective options for accessing Windows 3.1, 95/98 and NT eventually.

There is still a sense of catching-up as software writers find ways of presenting their products in more and more unusual visual styles. They sometimes use non-standard methods which create difficulties for the screen reader. It is then up to the developers of screen reader software to catch up once again.

Other people can also have difficulties with GUIs. The system which was intended to be intuitive can cause problems for people who do not cope well with visual presentations or who grew up with command line systems.

Some of the methods to enhance accessibility to CBI are similar to the ones used for Web accessibility, while others are specific to educational software. Although we will mention all of them, special attention will be paid to those which apply specifically to CBI.

## 17.3 POSSIBLE SOLUTIONS

Some form of adaptation will be necessary for developing an operating system, application software and learning materials which meet the needs of almost all people. Examples include screen readers with speech or Braille output for people with visual impairments and a host of alternative options for people with physical disabilities. However, much needs to be done if this goal is to be realised.

### 17.3.1 Icons

Icons are a quick and easy way of conveying information for many people and can be an effective medium for CBI (for example, back, forward, next), provided that the graphic is well designed. However, these symbols often do not convey the intended meaning and there must be an alternative method of performing this operation to highlighting a graphical representation. For screen readers, icons sometimes present problems. Current screen readers are able to recognise many standard icons and users can use screen reader facilities to label additional ones which are encountered in some software. However, labelling may require sighted assistance. Large expanses of graphics make software inaccessible to people who have severe vision impairment. One example might be to ask the user to click the mouse on part of a map to obtain information about that region. If no alternative method was offered, the product would not be usable by a blind person.

Alternatives include 'shortcut' keys—that is, selecting a function by pressing a key or key combination instead of clicking the mouse on an icon. A third alternative is to select a function through a menu system. A well-designed menu structure can be a very effective interface between the user and the software. The desired menu can be pulled down and the required item can then be selected by mouse, arrow key or shortcut key.

As an example, Andrew (one of the authors of this chapter, who is blind) and his sighted 14-year old son have different methods of using WordPerfect. If Andrew's son uses it after Andrew has finished, his son immediately turns on the toolbar. This allows him to select required items with minimal difficulty with the mouse. On the other hand, if Andrew uses it after his son, Andrew immediately turns off the toolbar to avoid clutter and to stop his screen reader chatting unnecessarily. An Internet site from another educational institution in Australia is designed so that students work on-line. However, it is very difficult, if not impossible, for people using screen readers to navigate, as the site uses pictures without 'alt tags', which make it possible for screen readers to read out what the picture contains. The same course is available in print from the Open Training Education Network (OTEN), which provides notes in electronic format to people with visual disabilities. However, thoughtful design of the Web site, meeting guidelines set out by the World Wide Web Consortium, www.w3c.org/wai, would have allowed universal access to it (see Chapter 10).

Flow charts are difficult to process for some people with acquired brain injury from an accident. Therefore, some learning materials making major use of flow charts may be difficult for these learners if alternative formats are not provided. However, learners with dyslexia or other difficulties in processing large amounts of text may find flow charts very helpful.

## 17.4 THE INTERNET

This has much to offer people who are often not considered when new interfaces are being developed. As the Internet has become more popular and commercial interests have become more involved, the risk of excluding minority groups has increased. It is completely feasible to gain access to the Internet with a very modest computer. However, as the user friendliness of the Internet has increased, so has the reliance on graphical presentations. This jeopardises access for those with visual impairment and those using more modest computers unless they are prepared to use text-based software. Finding an Internet provider who offers this access can, however, be difficult.

### 17.4.1 Text labels

Many Web sites are constructed in such a way that a person relying on text cannot make sense of them, often because of the graphical presentation. What can be done to make Web sites accessible to just about anyone? Using a graphic as a link without any accompanying text can present enormous problems. People using text-based browsers do not know where the link will take them without actually going there. People using screen readers, with either a text browser or a GUI, will be in the same situation. So it is recommended to use a text label on each link.

### 17.4.2 Unnecessary graphics

A trend among Web site developers is to take trouble to make the site as visually appealing as possible. This can, if not done carefully, effectively exclude access to the site by many people. One Web site dedicated to promoting graphic art, for example, would be expected to be rich in graphics. However, if a site provides more general information, it can easily antagonise visitors by heavy use of graphics which take time to download. So graphics should be:

a. used in context and not for their own sake, and

b. accompanied by text labels.

When teaching art history, for example, you might include a paragraph beneath each picture with a description of colours, tones, light and shade, shapes and their composition in the picture. Not only does this give visually disabled visitors a feel for the picture, but it also teaches sighted visitors about visual analysis.

### 17.4.3 Tables

These can pose problems for people using screen readers. Consider a different method of presenting the material or, even better, offer the user the option of the table or an alternative presentation.

## 17.5 LANGUAGE USE

What is read should be written in simple, clear language—this seems rather obvious but some technical writers tend to become bogged down in difficult language. Where people need to become familiar with technical vocabulary, this can be defined and experiential examples given. For some people, it is difficult to learn technical language after one read, so designers should consider giving a variety of vocabulary-building exercises and practice in the use of technical language through the materials.

When used in an individualised or self-paced mode, learning materials may need to be written more as a 'learning conversation' (Baåth, 1982) rather than in a formal, textbook style. Some textbooks use such language as "The reader may find it necessary to . . ." or "It is important that consideration be directed to . . ." The reader can be directly addressed by the writer and the grammatical structure simplified: "You'll need to do this because . . ." or "You'll find it's important to do this as . . . " These are more easily understood and more conversational.

People with reading difficulties will profit from the use of diagrams and clear layout, with a reduced amount of text, to highlight important concepts and make it easier for them to navigate through the materials.

## 17.6 ASSESSMENT

Worldwide, competency-based assessment is being increasingly used in many education and training areas. That is, assessment which measures the learner's achievement of competencies, whether these involve cognitive, attitudinal or performance skills. Designers need to consider carefully how competencies should be measured for people with disabilities (although assessment for some skills will be out of the question for some disability types, for instance, people with visual impairments and flight navigation).

Assessment events and criteria may already have been specified but the following questions still need to be considered.

- How are these skills going to be used in a job?
- Therefore how should they be measured?
- Can the broadest range of people achieve these skills or does the assessment criterion need to be changed?

Ensure that no one will be disadvantaged by the assessment method chosen. If possible, provide alternative options to cater for different learning styles and disabilities. Oral work as well as written work needs to be considered. When learners are working on computers, consider group assignments with an oral presentation. Assessment only on computer may not achieve all the competencies required. Collaborative learning should be acknowledged, as well as competitive learning (see Bryant and Rivera, 1995, for using assistive technology to accomplish this; Goor and Schwenn, 1993, on how to do it). Some trainers and teachers may feel that group assessment is cheating but, on the job, many tasks need to be accomplished in groups.

Consider learners' geographical restrictions. For example, people in one course may be required to walk around a block and draw a map of it in their local area. Prisoners would be expected to have difficulties as they would not be permitted to walk in the local community. So when devising assessment activities, whether for computer-based instruction or other forms of instruction, consider the variation in your target audience.

Developers sometimes have difficulty determining the difference between assessment for diagnostic purposes and summative assessment—that is, whether you are attempting to find out what your learners know in order to remedy their deficiencies during a course, or whether it is the end of a course and you are producing a 'final report' on what has been achieved. Ensure the difference is understood and specified.

In some pre-vocational courses, people may not have studied for a long time, and may not have been successful in their previous attempts at learning. In this case it is essential to make the first assessment task/s easy to achieve and not too time-consuming so that people can build their confidence as they progress. But at the same time, endeavour not to be patronising! Many sugary comments in learning programs or comments which are not well thought out are a real turn-off.

Quirk (1994) discusses issues in language and literacy affecting competency-based assessment (see Quirk, 1994, p. 36 for a summary). Disabilities Services in NSW, Australia (1994) has written a report on reasonable adjustment to competency assessment mechanisms for people with disabilities, giving examples of how this might be done (see Quirk, 1994, p. 47, section 4.3).

One of the most important points the report makes is that "there are very different disabilities and different degrees of disability" (Quirk, 1994, p. 32). They state in their summary that "the differences within the group classified as 'people with disabilities' are actually more extreme than any differences between the group and other members of the population" (Quirk, 1994, p. 34). You will find, for instance, that people with severe visual disabilities will usually require text presentation, while people with severe hearing impairment benefit from pictorial rather than text presentation. People with acquired brain injury benefit differentially from pictorial representation—the brain may have been injured in different places with differing severity.

## 17.7 TEXT SIZE, STYLE, FONT AND STRUCTURE

Designers should consider the consistency of font style and size. You may have seen pages on the Internet which are difficult to read for everyone from the point of view of background and colour! Broad fonts (e.g., Arial, Helvetica or Universal) are easier to read than narrower ones (e.g., Times, the default on Word) which are more difficult to read.

Always give an overview of the learning, either in a decision tree, flow chart, concept or mental map. Mind map© (Buzan and Buzan, 1993) has been registered as a trademark. People can build a map of what is coming and then fit it into a structure. Two Web sites contain colour illustrations of idiosyncratic mind maps. www.sharedvisions.com/explore/literature/mindmap.htm provides us with an example from Tony Buzan as used in his two-day lecture series. The site at http://ourworld.compuserve.com/homepages/marco_cosimetti/link2.htm has a more pictorial example with links to different sites used by the author. You will also find many examples of mind maps in Buzan and Buzan (1993). However, note that decision trees cannot be read by a person using a screen reader—also include a textual description.

To build a map, begin with a central concept—say food—and from there draw lines out to all the ideas that spring to mind. You can then subdivide these further (meat, vegetables). However, note that these cannot be read by a person using a screen reader—also include a textual description.

You can come up with non-examples (where do milk, cheese and eggs fit, for example) and different possible classifications (vegetables by colour or genus/species) depending on the purpose of the classification. You will find that, as you analyse your original diagram, you can revisit and elaborate it. When you read a learner's map, you can tell how far the learner's knowledge has progressed and how much effort they have put into the exercise!

Ensure the structure of the materials is clear. Remember that people with physical difficulties have trouble with the physical task of scrolling through text. Some people with neurological impairment, including acquired brain injury, have trouble with diagrams and flow charts—as do others with visual and intellectual disabilities—so alternative formats are helpful.

Consider how people will navigate through materials and ensure that the structure is standard from section to section, to help them predict in advance. If you include learning outcomes, flag them so they can be easily found. Cut down, as much as possible, on referring backwards and forwards as this becomes tedious and irritating.

Use practical and concrete examples of why things are done and how they will help on the job. Such explanations can also assist motivation: if you know why you are learning, it is easier to learn.

Use graphics and diagrams plentifully. However, text should be able to stand alone. People with vision impairment need a verbal explanation as well. If the diagram is essential, see if it can be simplified so it can be generated as a raised line drawing to complement your computer-based program. Provide a sample ground plan if people are asked to use it or provide a picture if they need to generate one themselves.

People with hearing impairment often miss information from radio and TV programmes without subtitles. People with literacy gaps miss written information. So when you need to refer to current affairs or background information from newspapers, magazines or TV programmes in your learning programme, either give the information where it is referred to or ask the people to find it, giving them references.

Make sure you check the students' understanding of the text explanation—people with hearing impairment and literacy gaps may assume they have understood but may not learn in sufficient depth to be able to use the information later. This can apply to everyone.

Repeat content in different ways and simplify it, but make sure repetition does not become boring for those who understood it the first time. State the concept, give information on how to learn it, then give the conclusion, stating the concept again.

Use cues to assist in recognising and using the structure. Consider emphasising it graphically by using different fonts and/or sizes for section headings to those in the text. For example, OTEN–DE uses Helvetica 18 point for A (or main) headings and Helvetica 14 point for B (or secondary) headings with appropriate spacing.

Use summaries at regular points throughout the topic, preferably in different ways (diagram, flow chart and table) to meet different needs including those of people with learning difficulties. However, note that these cannot be read by a person using a screen reader—also include a textual description.

Better still, ask the user to summarise, giving a possible structure. Give positive and negative examples to illustrate what does and does not belong to a concept, for instance democracy. This is a very abstract and elusive concept for some people who do not have the wealth of background information. Often developers overestimate the knowledge and experience of people, particularly those with disabilities.

## 17.8 MULTIMEDIA

Videos can be used as supplements in CBI, rather than as essential content. As indicated above, people who are blind cannot learn from video without explanatory audio. Those with hearing impairment need subtitles—these are preferable to signing—as more people with hearing impairment can understand them. Ensure that all essential video clips are subtitled.

Hofmeister (1992) gives guidelines on developing video for a diverse group. People need to ask themselves questions before they view the video and revise key concepts after viewing. Consider asking people to formulate their own questions. Many people with disabilities are not as confident about using technology. It may also take a long time to download programs with multimedia.

When using audio, make sure the voice is deep and slow so it is easily heard and understood. Female voices can also be deep, so try to ensure both are included. People with neurological impairment may have difficulty with visual presentation because their visual memory may be distorted—they have trouble extracting meaning from text and graphics. However, they have less difficulty with aural comprehension, so some will do better with audio.

The Web is very much part of the multimedia era. Such software can be very appealing and, in the right context, useful. It can be even more useful with some thought to its preparation. Both audio and video clips can be attractive and interesting. However, they are not useful to deaf or blind people, respectively. Frustration can be avoided by providing text captions of the clips and voiceovers/explanatory audio for the video.

## 17.9 LEARNING TO LEARN AND BRIDGING

Learning skills should be integrated in the content. You can include tasks which model how an expert would deal with the content. You could also include information for learners about strategies of memorising particularly difficult technical terms or sequences of steps in a process. Also consider whether some learners may need basic bridging skills to acquire content in particular areas. In electrical areas, for example, there may be a need to revise algebraic concepts for the mathematics content in the materials. Metacognitive strategies for people with disabilities are dealt with by MacMillan *et al.* (1986), particularly research on learning and cognition (MacMillan *et al.*, 1986, pp. 696 onwards).

## 17.10 CONCLUSION

It is difficult to give blanket recommendations for learning materials and computer-based instruction for all disability types. Strategies which assist some people need to be carefully considered for others, for instance, diagrams and tables may create problems for people with visual impairment. Therefore, it is important that alternative forms are provided.

The potential exists to create operating systems, application software and learning materials which can meet the needs of very many people with diverse needs and equipment. It is also possible to create software which precludes many people from benefiting from it. It is now easier, in fact, to produce software which looks good than to make it universally accessible.

Universally accessible software does not necessarily mean that everything should be available to everyone in exactly the same way. For example, blind people may not—at least at this stage—be able to use a drawing program in the same way their sighted colleagues can. Similarly, audio clips cannot benefit deaf people directly. What is important is that software which could and should be accessible is not made difficult or impossible to use due to unnecessary reliance on one modality only.

A large part of the reason for some software not being user friendly to a broader range of users is that those who develop it are not aware of the types of issues discussed above. It is therefore important to do all we can to make them aware.

## ACKNOWLEDGMENT

The above is based on the experience of teacher consultants who deal on a daily basis with OTEN students with these disabilities and learning difficulties.

# Part 6

# The Future

## CHAPTER EIGHTEEN

# Telecommunications—Accessibility and Future Directions

Gregg Vanderheiden

### 18.1 INTRODUCTION

In the past, products were designed to operate in a rather fixed fashion. Their design was also targeted at people in the center, or high center, of the ability curve. That is, products were (and usually still are) designed for people who either have all of their abilities or actually have slightly above average abilities. People who have disabilities typically gain access to these products by either purchasing special versions (where they were available) or by attaching assistive technologies or other special interfaces.

A number of things are changing, however, which will dramatically affect the future of product design and accessibility. This chapter outlines some of these changes and describes some current and future technologies that demonstrate this new flexibility and potential for built-in accessibility.

### 18.2 TECHNOLOGIES USED TO CREATE NEW PRODUCTS ARE DIFFERENT FROM THOSE OF THE PAST

In the past, products were more electromechanical in nature. This meant that if a company wanted a product to behave differently, it had to be physically redesigned differently. For example, telephones used to be electromechanical and their interface was determined by the design of the mechanisms. Today, almost all products are controlled by a microprocessor running a program. The interface of today's telephones, for example, is defined by the software program running in the phone itself. As a result, it is easy to cause phones to behave differently by simply changing the instructions in the program. It is also possible to have telephones perform different tasks for different users, or for the same user in different situations, by allowing the user to select different behaviors.

With the speed of processors increasing and the cost of processors and memory continuing to go down at a steady rate, it is possible today to build flexibility into the interfaces of products in ways that were unheard of before. The power that will be available in small products tomorrow is staggering. The Nintendo game machine of a few years ago had the processing capacity of a Cray supercomputer in 1985—and those Nintendos are now slow and obsolete by today's standards. This increasing computing power is steadily making its way into even simple products, making them more flexible and adaptable to different situations and users.

At the same time, interest in mobile computing and telecommunications is providing additional incentive to create products with more flexible interfaces. Individuals want technology that allows them easily to change the way they interact with a product based on different needs and situations. For example, people want to access information while at their desks; while walking (no standard keyboard); while driving (eyes busy); while performing tasks that keep the hands busy; while in noisy environments (can't hear); or while in quiet environments like libraries or meetings (can't use sound). Additionally, the rapid advances in voice technologies are making verbal (e.g., words) control of products, and the use of natural language to operate products, an increasing reality.

These trends are coming together to provide all users with systems that will have unprecedented flexibility and adaptability to different needs, at different times and in different environments. This same flexibility will have profound effects on the ability of products to adapt to the different situations faced by people who have temporary or permanent disabilities.

This chapter will look at the future of telecommunications through a series of scenarios and stories that begin with today and extend into the future, providing a picture of potential telecommunication uses as they relate to the future needs of people with and without disabilities.

## 18.3 SILENT MESSAGING—TTY FEATURE IN DIGITAL CELLULAR PHONES

When digital cellular phones were introduced, compression algorithms distorted the TTY (originally meaning TeleTYpe but now used for text telephones) signals, making it difficult, or impossible, to transmit clear TTY messages over digital cellular phones. To address this problem, the telephone manufacturers changed the software in the phone so that it would recognize the TTY tones and send them as data instead of sounds. The data stream is then used to reconstruct clean TTY tones in the network at the same point where the compressed speech is reconstructed. The result is a very clean TTY signal.

Looking closely at this situation, we see that this means that these digital cellular phones are capable of sending and receiving TTY characters as data. With little additional effort these standard, off-the-shelf phones could be programmed to present the TTY characters on the phone's display as they are received, thus providing a built-in TTY receive functionality. Individuals who are deaf, but who can speak, could then use these standard phones without any special modifications and without needing a separate TTY.

If the phone happened to also have some way of entering alphabetic characters, it could also be used by individuals who are deaf and cannot speak. Again, the TTY characters coming into the phone could be displayed. To respond, the character entry mode (keyboard or keypad, etc.) could be used to type characters to be sent out as TTY codes. The software to do this would be quite small since it would just involve connecting the key entry routine to the TTY data transmission routine. This capability would allow these phones to act as TTYs themselves.

This feature might be useful to many people, not just individuals who are deaf. Phones could be designed with a 'silent operation' mode that would allow users to communicate quietly in text during meetings, in libraries, lecture halls—anywhere a person would not be allowed to talk but might have a need to communicate. Where this feature is implemented, individuals would be able to purchase a regular cell phone, or borrow one, and be able to communicate with it in text as easily and naturally as others who use the phone by voice. In the future, TTY and text chat of many different types may be interchangeable.

## 18.4 CAPTEL—CAPTIONED TELEPHONE CONVERSATIONS

This second scenario deals with a service called Captel, short for *captioned telephone*. The Captel technology is already available today in limited markets on a pilot basis and soon will be rolling out nationwide across the United States. To the user it looks like this:

> *Mary can hear, but she is getting older and doesn't hear very well. She enjoys conversations with her children but often misses what they are saying and just has to fake it. Occasionally, this has created real problems since she did not pick up critical information that was given to her by her doctor about medications. She also recently missed a family gathering, because she did not understand the call from her daughter telling her that the time and place had changed.*
>
> *As a result, Mary was delighted when her friend described the Captel telephone. This telephone looks like a standard telephone except it has a small caption button in the lower-right corner next to the dial pad. It also has an LCD display across the top, above the keypad, like many other telephones do. Whenever making a call, Mary now pushes the caption button and then makes her regular phone call. In fact, everything about the phone call is the same as it was before, except that whatever the person at the other end of the line is saying appears as text in the display at the top of her phone. Thus, she can carry on a two-way conversation with her doctor or children, etc., and both hear and see everything they are saying. She can hear the sadness, or joy, or concern in their voice just like before. But for the first time in a long time, she understands everything they say and never misses a word since everything they say is displayed on screen in nice large letters as they talk. And the nice thing about Captel is that it works with anyone's voice with perfect recognition.*

Some day this technology will occur through the use of computer-based speech recognition. Today, speech recognition is not nearly good enough and does not accommodate a variety of callers, including people with accents. So, instead, the Captel system operates by having a third person (a Captel operator) listen in on the call. Pressing the caption button before dialing links the operator automatically into the call. The operator listens to everything that is said by the other party and simultaneously re-speaks the same words very distinctly into a computer that provides a voice recognition function. Because the operator does this all day, she has trained the computer to recognize her voice and is able to get a recognition rate that is extremely high. In addition, whenever the computer makes a mistake, it can be seen onscreen and corrected before it appears on the display of the Captel telephone. For more information see Ultratec's Web site at www.ultratec.com

## 18.5 THE 'TRY HARDER' CONCEPT

The 'Try Harder' concept is a concept to allow a natural transition between today's human-assisted approaches and tomorrow's fully automated services. Applied to the area of speech recognition, it might look like the following: An individual would like to use fully automatic, computer-based speech recognition. However, today this would only work in certain circumstances and with certain speakers. If it turns out that the person being called has a heavy accent or, for some other reason, the automatic speech recognizer does not work, adding a 'try harder' button on the product could allow a user to be able first to try the automatic (and much less expensive) speech recognition to see if it will work. If it doesn't, the person simply pushes the 'try harder' button and a human being is alerted to view the output of the automatic speech recognition and make corrections. Pushing the 'try harder' button again could cause it to shift to a situation where a human being is re-voicing, using short-hand or some other technique to provide more accurate speech-to-text translation. Each press of the 'try harder' button could bring in more skilled individuals at a higher cost.

With the 'try harder' technique, systems can be put in place to work today that will naturally migrate towards and incorporate newer technologies (in this case speech recognition) as they become available. This will provide a faster, lower-cost system over time. By developing systems in this manner, we can create the services and the infrastructures that are needed today and will need to be in place so that advances can be automatically incorporated as they become available.

This 'try harder' strategy can be used not only with speech technologies but with other technologies as well, as is described below.

## 18.6 LISTENING PEN—PERSONAL INTERPRETER

A more general form of the speech-to-text concept has been described over the years by multiple researchers and prognosticators (e.g., Apple Computer, 1989) to provide deaf users with a personal, portable interpreter. One example (Vanderheiden, 1995) took the form of a hypothetical device consisting of a small, directional microphone that could have the appearance of a pen, accompanied by a small, alphanumeric display that could be built into a pair of glasses. A deaf person could carry the device inconspicuously and, whenever a conversation occurred, simply angle the pen toward the mouth of the person who is speaking. The speech could then be picked up and transmitted to a central location where it could be transcribed into text. The transcribed text could then be sent back to the user and presented in the display built into the glasses. This could be projected virtually so that the text actually appears to float in space about 2–3 feet in front of the user. In this fashion, the user can look directly at the person who is talking and literally see the words written across the person's face. Therefore, individuals who are deaf could have the ability to carry on face-to-face, or eye-to-eye, conversations with anyone who might be talking to them. As technology advances, the central 'transcribing' service could be more and more automated and eventually be entirely automatic. (In fact, in the far distant future, the processing power to do perfect recognition may fit into the pen itself.) In September 2000, Ultratec (www.ultratec.com) developed a new service called 'Instant Captioning'™ that makes this general concept a reality. Currently in field-testing, Instant Captioning will provide a capability similar to that described here in different technology formats.

## 18.7 REMOTE SIGN LANGUAGE INTERPRETATION

Remote sign language interpretation would operate in a fashion similar to the listening pen and Instant Captioning. However, instead of text being sent back, the image of a sign language interpreter is projected to the user. Specifically, the person who is deaf would see a sign language interpreter floating in space, communicating to the person what is being said by another person. It may also be possible for a person who is deaf to sign to a small camera and have the signing sent to an interpreter, who could voice the message, and send it back through an audio device to the deaf person's communication partner. In this way, the person who is deaf would be able to carry on a face-to-face conversation with an individual who could not sign or understand sign language, even if the person who is deaf could not speak or understand speech.

Initial versions of this have already been implemented experimentally using a live person doing the sign language interpretation from a remote location (Barnicle, 2000). The image of the signer was sent over Internet 2 and a high-speed local network and projected onto a screen in the front of the room as well as onto a head-up display. Although sending a video signal with enough resolution and reliability to be used for sign language and finger spelling is not possible on the Internet today, it is possible on Internet 2 and should be on future internets.

The need for a high-speed connection, however, can be removed by using an avatar (computer-generated person) as an alternate approach to sending a complete video signal. The avatar's movements could be abstracted from a real (live) signer, or they could be generated via direct audio-to-avatar translation. The abstracted movements could then be sent over a lower bandwidth connection to control an avatar at the user's end. The avatar would, in turn, present the sign language interpretation on a personal display. Research is currently underway on both text-to-sign language interpretation and the generation of software avatars that can carry out the sign language gestures with sufficient accuracy and fluidity.

The potential impact of a personal interpreter system as described here would be tremendous. Most people who are deaf cannot afford to have an interpreter with them at all times. Yet, without one, many are unable to communicate and interact with other people that they meet on a personal, professional or consumer basis in their community or businesses. Personal interpreters would allow individuals who are deaf more freedom in communicating and the ability to call upon the services of an interpreter only when they are needed.

Remote interpretation could also boost productivity levels for interpreters by allowing them to work out of their homes on a demand basis, thereby greatly increasing the number of interpreters available, while decreasing interpretation costs for people who have disabilities.

## 18.8 MORPHED SIGNERS AND TALKERS—VIRTUAL COMMUNICATION SKILLS

These concepts can be extended further when individuals are involved in communications over telephones. Using morphing technology, it is possible to take an image of a caller and combine it with a signing avatar so that one ends up with an image that looks exactly like the caller doing the signing (even though, in fact, the caller is just talking). In this way, it is possible to send an image of someone signing over the picture phone even though the person does not know sign language at all.

Similarly, it is possible to take an image of a person and morph their lips so that it looks like they are talking even though they are not. Thus, a person typing away on a keyboard could appear to be sitting there calmly talking, lips moving, and speech apparently coming out of the person's mouth, even though the person is not actually talking.

Combining these two applications could result in the following scenario: An individual on one end of a videophone call who is deaf and cannot speak appears to speak to others on the phone. When they communicate back to him, he sees them signing even though they are not. Eventually, gesture recognition will be so good that the individual who is deaf could actually be signing and seeing the other person sign back, while the other person would talk and see the deaf person talking back. In this scenario, each individual can choose to either present himself or view the other person in the form that is easiest to interact with.

## 18.9 ASSISTANCE-ON-DEMAND

In the above examples, the concept of an 'assistant-on-demand' was introduced. Basically, an 'assistant-on-demand' is an individual who could be called upon to assist someone with a disability any time it is required but who would not be around the rest of the time. In the examples above, the assistant was an interpreter helping someone who is deaf communicate in an auditory environment. However, assistants-on-demand could also be used by individuals with visual or cognitive disabilities.

For example, someone who has low vision or blindness might find it very convenient to be able to use a small camera and a video telephony link to allow them to access an individual who could act as a vision assistant as needed. Although there will soon be small portable devices that allow people who are blind easily to read any text in their environment, it will be some time before there is anything automated which can deal with a complex visual scene. However, a person who is blind could easily use a 'visual assistant-on-demand' (VAOD). For example, an individual may be sitting in a lecture listening to the professor. Eventually, the professor may refer to a chart or diagram. At this point, the individual who is blind could activate the VAOD and instantly have someone help them interpret the chart through verbal description. The video device held by the individual could even allow the remote viewer to pan and zoom better to aim the camera or zoom in on the chart.

The VAOD could also be useful during video teleconference calls. Again, the individual who is blind may be able to participate in a teleconference call with little need for a visual assistant. If one of the participants, however, began using visual aids, the individual who is blind could simply press a button to bring a VAOD into the teleconference. Because the entire teleconference is electronically mediated, it is easy to keep a 30 or 60 second loop running. Thus, when the VAOD is brought into the conference, the assistant can actually begin viewing starting 60 seconds or so in the past. In this fashion, the person who is blind can have the assistant describe the original visual event that caused them to call the VAOD in the first place. In order to keep the assistant's descriptions from bothering the other participants on the call, the assistant could be connected in such a way that only the person who is blind hears the description.

Another technique, which might be called 'freeze and catch up', could also be implemented. In this scenario, the video conference could be paused only for the person who is blind to allow the video exhibit to be described to him or her. When the blind person finished having the scene described to them, they could view the conference call starting at the point they 'froze' it. They would, however, view it at a slightly faster than normal speed so that they could catch up to the real time discussion. In this fashion, an individual who is blind could freeze the action in order to get verbal descriptions, and yet not disturb the teleconference, nor miss any of the discussion.

## 18.10 THE COMPANION—A COGNITIVE ASSISTANT-ON-DEMAND

An example of a hypothetical device that could provide assistance to individuals with cognitive disabilities was included in Rowitz's *Mental Retardation in the Year 2000* in a chapter entitled "A Brief Look at Technology and Mental Retardation in the 21st Century" (Vanderheiden, 1992c). The chapter described 'companion' technologies that assist people with cognitive disabilities by both extending or enhancing the individual's cognitive abilities, and also augmenting the user's cognitive abilities with a second separate cognitive entity (artificial or real). Aids which incorporate artificial intelligence techniques to the point that the devices begin to take on the characteristics of an intelligent entity themselves could be very helpful to an individual with mental retardation, and would fall into this category.

One example of such a device was a hypothetical aid called the 'Companion' which combined both artificial and real personal assistance.

> *The Companion consists of a small device approximately the size of a large wallet. It has four or five large buttons on it, which are brightly and distinctly colored and have symbols on them. One of the buttons stands for 'Help'.*

*Two other buttons stand for 'Yes' and 'No'. Another button is a request button. The Companion has voice output and speech recognition. It has an artificial intelligence system programmed within it which is specifically designed to facilitate problem-solving and crisis resolution and the Companion acts as a reminder and monitor system for the individual. The Companion has a built-in GPS system which allows it to keep track of its exact position using signals from navigation satellites. Finally, the Companion has a cellular communication system similar to a cellular telephone, allowing it to put the individual into instant contact with a crisis line in case of an emergency which cannot be easily handled by the Companion.*

*In daily use, the Companion would work something like this. Certain regular activities would be programmed into the system. The Companion could wake the individual up in the morning and then periodically ask questions to help cue or prompt them through their morning routine, or operate as a check for individuals who do pretty well at progressing through their morning routines on their own. While of some benefit in the routine operations, the Companion would often be more useful in helping to remember the breaks or changes in their routine. This would include days where the individual is not supposed to go someplace they normally would (due to a special appointment, a holiday, etc.), as well as unusual things they must do or places they must go (doctor's appointments, etc.). Because the system always knows the physical location of the user, it can automatically discover when suitable progress is not being made toward the programmed goal. For example, if the individual has a doctor's appointment, the Companion would remind them in time for them to get to the doctor's office. If the time for the appointment was drawing near and the individual had not made progress toward the doctor's office, the Companion would provide additional reminders to the individual, help them problem-solve the situation or put them in contact with someone who could help them out. As the day progressed, the individual would be able to request information or assistance from the Companion by pressing the 'Request' button. The Companion would respond by talking to the individual and running through a list of functions to see what it was that the individual wanted. Because of the speech recognition capability of the Companion, the individual could also simply respond to the initial questions of the Companion by saying what it was that they wanted. Some of the functions that the Companion could provide include providing time of day, providing directions, checking appointments, providing telephone numbers, providing addresses, etc.*

*For example, Tim falls asleep on the bus on his way home from work and rides past his normal stop. When he wakes up, he looks out the window and finds himself in a totally strange part of town. He panics, gets off the bus and begins walking aimlessly about the streets, becoming more alarmed as he goes. The Companion, recognizing that he is in an area that he has never been to before and that is not on his agenda, beeps and asks him if he is okay and if he knows where he is going. Using speech or the 'yes/no' buttons, Tim answers the questions until the Companion is fairly certain he does or does not know where he is headed. If the Companion determines that he has a problem, it would go into crisis/problem resolution mode, figure out what the problem is, and attempt to help him.*

*The Companion would start by asking questions to determine what type of problem exists. It would then run through a number of problem-solving strategies to try to help Tim solve the problem himself. (While doing this, it would be functioning in an enhancement mode.) If this did not work, then Tim might be advised to seek help from those around him (depending upon the environment and situation). Finally, if the Companion was unable to help Tim solve the problem himself, or with the assistance of those around him, the Companion would use its wireless phone-like capability to contact a central, shared 'Help' facility. This central facility would be manned 24-hours per day by individuals trained to provide assistance and problem solving to people with the Companion device. Since the central facility would have a basic file on each individual, as soon as the Companion contacted it, background information on Tim would be instantly displayed on the operator's screen, along with any information that the Companion had been able to glean through its processes, including the individual's current location. At this point, Tim would be in direct voice and visual contact with the operator, who would be able to talk to him in a fashion appropriate to his abilities, determine what the problem was and help him to resolve it.*

*Tim does not have to wait until the Companion detects a problem if he is in trouble. He can directly signal that he has a problem by pushing the Help button. Tim could also skip the Companion's assisted self-resolution phase and contact the central operator right away by pressing the Help button more than once.*

*The Companion would also have a viewfinder as is found on a camera. Whenever Tim needs help understanding what is written on a sign or piece of paper, he could just look through the viewfinder, aim it at the text and push a button. The Companion would take an electronic 'picture' and then read the text aloud to Tim. This feature could also be helpful in the way-finding strategies because the Companion could determine what the sign says (if it were a familiar type), and provide assistance. For example, if Tim was near a bus stop, the bus numbers could be noted and used to help him get on the correct bus. Similarly, when a bus approaches, Tim could just aim the viewfinder at the words or numbers written on the bus and the Companion could confirm whether it is the proper bus for Tim to board. This reading capability might also have a therapeutic learning value to Tim since it would display the picture of the text and then highlight the words one at a time as it spoke them. In this way, the Companion could facilitate Tim's learning and recognizing common or familiar printed words. It might also be possible for the Companion to transmit the digitized picture back to the central operator in the event of an emergency or problem-solving situation. As a result, the operator could 'see' the problem more clearly by simply asking Tim to 'take pictures' of what the operator was interested in.*

The ultimate purpose of the Companion would be to allow individuals with mental retardation to live more independently. If the Companion could enable an individual to live safely in a less supervised or more independent fashion, the cost savings would quickly cover the cost of the Companion. Even a moderate shift in independent living status can produce savings of $10,000 or more per year. The cost of the Companion could be quite reasonable through widespread application with elderly individuals, where it could help allow them to live safely outside of nursing homes for longer periods of time. In nursing homes, it could allow greater degrees of freedom while still permitting maintenance of supervision to whatever level was desired.

Technologies such as the Companion are easily within our reach. GPS technology, cellular telephones, miniature cameras, voice synthesizers and microprocessors are all in use today. The artificial intelligence and expert system routines that would be needed for the problem-solving systems in a Companion-class device exist but not to the degree of sophistication required for this application. However, steady progress is being made in these areas, and the computing power necessary to support these capabilities is advancing rapidly and will easily outstrip our ability to apply it effectively. What is not known is how to best utilize these technologies, how to provide useful prompting to help a client figure out a problem rather than thinking for them, and how to identify problems without a visual context.

## 18.11 WINDOW TO THE WORLD ('WINDOW')

The final scenario in this chapter is a description of a hypothetical multi-purpose communication and information device in the not-too-distant future. All of the techniques and technologies described in this product exist today in laboratories. They just have not been combined in this way or deployed to the field. Also, the form factor would be larger today.

This product is presented here because it shows how it is possible to design products in the future that could provide full cross-disability access (that is, a single product that can be used by people with a wide range of different disabilities) as a part of the product's natural design and standard mass market features. To help demonstrate this, the product will first be described as it would be used by people who do not have disabilities. Its use for people with disabilities then follows:

The 'window to the world' ('Window') resembles a thin pocket-sized device, measuring 3.5 inches x 4.5 inches, which is ¼ inch thick when closed. It opens up much like a man's wallet to measure 3.5 inches x 9 inches, which is ⅛ inch thick except for a ½ inch strip down the right edge which remains ¼ inch thick. This thicker strip contains two cameras, a microphone and controls for the device (see Figure 18.1). When open, most of the device is a large touchscreen interface that covers the entire inside of the unfolded device, with the exception of the aforementioned ½ inch strip down the side. The device has four small buttons—two each on the extreme bottom sides that are used in addition to the touchscreen to operate the product.

**Figure 18.1** 'Window' hypothetical communication and information device

With the 'Window' a person can access any information they have permission to access anywhere in the world—any book that has been published; any movie; any document; any report; all of their own personal information, files, letters, records, report cards; lunch menus; movie times; product prices or instruction manuals for products. None of the information is stored in the device. In fact, the 'Window' has no information stored in it. It is simply a window to all of the information that is stored on any device or service connected to the I-Net (a future more comprehensive and secure version of the Internet). One side benefit of this is that if the 'Window' is lost or broken, the user can just stop by any store and buy a new one (they would sell for an equivalent of $75), and they are instantly back in business with access to all of their information just like before.

The 'Window' would connect wirelessly to the I-Net. In most buildings and other locations (including cars) it would simply tie into the local network of the building (or car). Remotely, it would connect to a small device about the size of a pack of cards that boosts the signal and ties it into the I-Net via a cellular or satellite wireless connection.

In addition to providing access to any information, the 'Window' could also act as a telecommunication device. In addition, there could be network-based translation services available, making it possible to communicate across language barriers.

The device has two cameras built into it, one facing towards the user and one facing in the other direction—two cameras are used since it is cheaper and more reliable than a pivoting camera. It is, therefore, possible to carry out two-way videoconferencing. It even has a 'steady cam' feature so that the image remains stable.

Using the camera on the rear of the device, a person can use the 'Window' as a camera to take either still pictures or movies. Because of the high-resolution imager, there is a built-in digital 10x zoom capability. Since it does not store information locally, the device never 'runs out of film'.

In addition to using the 'Window' to take pictures or movies, the camera and zoom capability can be used as a high-power magnifier. This can be used for anything from removing a splinter to reading somebody's name-tag from across the room.

The device also has a capture-and-read function. For example, a person could zoom in on another person's name-tag and have the text captured and then read by the device. It could then use that text to look up information about the person in a database. It could also be used to capture the text on a sign when a person is traveling, for instance, and translate it or give location information using the built-in global positioning capabilities of the device.

Finally, the camera has an 'image stitching' function to read larger images. The user can just sweep the 'Window's' camera back and forth over an image and the device will stitch the image sweeps together to form a picture that is larger than the 'Window's' camera can take at any one time. For example, the 'Window' could be waved back and forth over a menu and then it would construct a full image of the menu. The menu could then be read to the person or translated into their native language. The 'Window' could also answer questions about the menu—for example, was a particular type of entrée listed on the menu.

Information can be entered into the device using either a keypad, which could appear on the touchscreen, or via voice recognition. When using the screen as a touchscreen, the user could specify the size of keyboard. When talking to the device, natural language can be used which is automatically translated by speech-to-text translation services both in the device and network as needed.

The 'Window' comes with a small earbud, which fits into the user's ear and communicates wirelessly with the 'Window'. The earbud acts as both a speaker (so the 'Window' can speak to the user or pass along a telephone conversation) and a microphone (to pick up one's speech as it comes through the Eustachian tube from the throat to the ear). This allows the 'Window' to hear a person more clearly despite background noise. Using the earbud allows the user to interact with the 'Window' in a completely hands-free and eyes-free fashion. One simply talks to the 'Window' and it presents back any information requested via the earbud. (If a person is in a very noisy environment or an environment where talking isn't permitted, then the user can simply pull out the 'Window' and use it in a completely silent/visual fashion.)

The 'Window' also comes with a head-up display that can be built into a pair of glasses or worn as a simple ornament that mounts along the side of the head. Both of these systems project an image onto the retina using a micro-miniature projector. This causes the user to see a display which floats in space about 2½ feet in front of them. This image is always in focus regardless of distance since the image is projected directly onto the retina. In this way, a person could be looking at any object or scene and still see the display of the 'Window' superimposed on it.

One of the great benefits of the 'Window' is that it is the world's best 'OnStar' device. Any type of help information can be accessed at any time from anywhere. And the built-in camera feature allows the person asking for help to communicate exactly what is needed or what the difficulty is. This is true whether the user wants to show the help site one's current situation, send them a picture of the device that one doesn't understand, show a diagram or chart that is confusing, or show a garden shop a strange mold or bug that is devouring the roses.

All in all, people who have the 'Window' find that it is a wonderful aid to almost any activity in their lives, including education, employment, recreation, health, safety and daily life.

### 18.11.1 Use of the 'Window' by people with disabilities

As discussed previously, designing systems to be truly flexible and adaptable enough to meet the varying needs of people who do not have disabilities will naturally result in products that are more accessible and usable by people with disabilities.

In the case of the 'Window', the product, even as described above, is naturally cross-disability accessible. For example, people who are blind can use the device in its 'eyes-free' mode. Any information presented on the screen can be presented to them via speech. If a person who is blind encounters any graphs or charts which cannot easily be rendered in text, a 'visual assistant' can be called upon to help describe the information to them. The built-in cameras on the device can also be used to access any text in the environment. By simply pointing the device, any text in that environment can be read to them. The zoom and image stitching capability of the camera can even allow a blind person inexactly to sweep back and forth over a printed menu and have the menu read to them. Again, if the environment or document is too complex for the device to handle, the user can call up an artificial intelligence agent or a human being to assist them at any point and time to deal with the visual task.

Individuals with low vision can use the built-in zoom feature either to magnify or to scan text and then have the text presented in large print on the screen. The information could also be read, whichever is more convenient.

Individuals who are deaf would simply use the 'Window' in its silent mode of operation, just as a person who is not deaf would use it in a very noisy environment.

In addition to having all information presented visually, a person who is deaf could also have any speech, either on the phone or in person, automatically translated into text and presented visually. This could be presented either as text or as sign language. In addition, the visual presentation can either be on the display of the 'Window' or, using the head-up display option, the text or sign language could be superimposed on the individual's surroundings—including the communication partner—or on a screen in the front of a room or wall. Using the built-in camera, it might even be possible for the individual who is deaf to use sign language, which could be converted to speech over the network and played from the 'Window'. In this way, sign language could be used to communicate with people who do not understand it but could instead hear the speech equivalent coming from the 'Window'.

Individuals who are hard of hearing can have a hearing aid built directly into the earbud. This way, the earbud could serve as a standard hearing aid and as an earbud to the 'Window'. In fact, wearing electronic earbuds would become the fashion since it would indicate that one was 'connected'. As a result, executives and others with hearing problems actually wear their earbuds (which have hidden built-in hearing aids) when they would not have worn hearing aids in the past.

People who are hard of hearing can also hand their 'Window' to the other person or ask the speaker at a lecture to put the 'Window' in the speaker's suit coat pocket. The user can then sit in the audience with their earbud and hear the speaker as if they are just 12 inches away (which they are from the 'Window'). If the lecture hall has a public address system, the individual can also simply tap into the signal using their 'Window' and have it shipped directly to their earbud. Finally, if the individual still has trouble understanding, the audio signal can be sent off, translated into text, and displayed on the person's 'Window'.

People with physical disabilities can use the 'soft' (screen-based) button feature of the product to enlarge the buttons to the size that works best for them. The 'hands-free' feature of the product could also be used simply to talk to the product. Finally, for those individuals whose speech is not very clear, a 're-voicing' service on the network could be accessed and a specially trained computer or human being could listen to the user's speech and simply re-voice it in a more articulate manner. This could help a person who has difficulty being understood by others and improve direct control of any speech-operated products or services.

Finally, individuals whose cognitive, language or learning disabilities make reading difficult could use the text capture and read function to have the information read to them. For some types of information, there could be services available to simplify the language used to communicate ideas for better understanding. This device could also carry out most of the functions described under the 'Companion' description earlier in the chapter. In addition, since the 'Window' is a thin client device and could eventually be quite inexpensive, there is less concern for the device being lost, stolen or broken. The cost to replace it is much less than the cost to provide its functionality in any other way.

Thus, the 'Window' not only provides built-in accessibility to all of the functions the product provides for people without disabilities, but also could act as a conduit to a number of special 'assistant-on-demand' functions. These functions could be designed to work specifically for people with disabilities but would, in all likelihood, be of benefit to all people who have difficulties at one time or another in accessing, seeing or figuring out things, people or situations.

## 18.12 GUIDELINES FOR THE FUTURE

So what does it take to create products in the future that will be cross-disability accessible and usable by people with functional limitations? The last scenario suggests that it is not a matter of adding special features for people who have disabilities but rather of designing a product that is flexible enough to allow users, who are operating under constrained conditions, to be able to use the standard products. These constraints might be just a noisy environment or the person may be hard of hearing. It may be because the person's eyes are busy (such as driving) or because they cannot see. Whatever the reasons, if the product is flexible enough and provides the user with enough options for input and presentation of information, then the future technology should be accessible.

The last scenario also suggests that our current strategies for making current technologies accessible may not make sense in the long run. Technologies in the future may be so different that the specific techniques or devices that we use to make products accessible today may not exist on future products.

Rather than looking to specific techniques to address today's technologies, we need to begin looking at general approaches for making products more flexible and easier to use in general. Table 18.1 provides a summary of what these general rules might look like. The rules are performance in nature (e.g., what to achieve) rather than being design guidelines (what or how to do it). The strategies are also general in nature but help to provide an understanding for the underlying principles.

The information in Table 18.1 is not a final definitive list but rather a first pass at providing a targetable summary of generic principles and guidelines which standard product manufacturers can follow in trying to create more flexible, usable next-generation information and telecommunication technologies.

How accessible future technology actually will be, will be a function of our ability to refine our understanding of these fundamental underlying principles and our ability effectively to communicate them to those who will be creating our future information and communication environments.

**Table 18.1** Basic principles and strategies for access to electronic products and documents

| Basic Access Principle | Why | How—General |
|---|---|---|
| Make all information (including status and labels for all keys and controls) perceivable<br>• without vision<br>• with low vision and no hearing<br>• with little or no tactile sensitivity<br>• without hearing<br>• with impaired hearing<br>• without reading (due to low vision, learning disability, illiteracy, cognition or other)<br>• without color perception<br>• without causing seizure<br>• from different heights<br><br>**Note**: other aspects of cognition are covered below | Information which is presented in a form that is only perceivable with a single sense (e.g., only vision or only hearing) is not accessible to people without that sense.<br><br>**Note**: This includes situations where some of the information is only presented in one form (e.g., visual) and other information is only presented in another (e.g., auditory).<br><br>**In Addition:** Information which cannot be presented in different modalities would not be accessible to those using mobile technologies, e.g.:<br>• Visual-only information would not be usable by people using an auditory interface while driving a car.<br>• Auditory-only information would not be usable by people in a noisy environment | **FOR INFORMATION:**<br>Make all information available either in<br>a) presentation independent form (e.g., electronic text) that can be presented (rendered) in any sensory form (e.g., visual–print, auditory–speech, tactile–Braille)<br>OR<br>b) sensory parallel form, where redundant and complete forms of the information are provided for different sensory modalities (synchronized), (e.g., a captioned and described movie— including e-text of both).<br><br>**FOR PRODUCTS:**<br>Provide a mechanism for presenting all information (including labels) in visual, enlarged visual, auditory, enhanced auditory (louder and if possible better signal-to-noise ratio) and (where possible) tactile form.<br><br>**Note:** this includes any information (semantics or structure) that is presented via text formatting or layout. |

**Table 18.1 (cont.)** Basic principles and strategies for access to electronic products and documents

| Basic Access Principle | Why | How—General |
|---|---|---|
| Provide at least one mode (or set of different modes) for all product features that are operable:<br>• without pointing<br>• without vision<br>• without requirement to respond quickly<br>• without fine motor movement<br>• without simultaneous action<br>• without speech<br>• without requiring presence or use of particular biological parts (touch, fingerprint, iris, etc.) | Interfaces which are input device or technique specific cannot be operated by individuals who cannot use that technique (e.g., a person who is blind cannot point to a target in an image map; some people cannot use pointing devices accurately).<br><br>**In addition:**<br>Technique-specific interfaces may not be accessible to users of mobile devices. For example, people using voice to navigate may not be able to 'point'. | Provide at least one mode (set of modes) where<br>a) all functions of the product are controllable via tactilely discernible controls and voice output of any displayed information required for operation including labels<br>AND<br>b) there are no timeouts for input or displayed information, OR allow user to freeze timer or set it to long time (5 times default or range), OR offer extended time to user and allow 10 seconds to respond to offer<br>AND<br>c) all functions of the product are operable with:<br>• no simultaneous activations<br>• no twisting motions<br>• no fine motor control required<br>• no biological contact required<br>• no user speech required<br>• no pointing motions required<br>AND<br>d) If biological techniques are used for security, have at least two alternatives with one preferably a non-biological alternative unless biological-based security is required. |

**Table 18.1 (cont.)** Basic principles and strategies for access to electronic products and documents

| Basic Access Principle | Why | How—General |
|---|---|---|
| Facilitate navigation:<br>• without sight<br>• without pointing ability<br>• without fine motor control<br>• without prior understanding of the content<br>• without the ability to hear<br>• without good memory | Many individuals will have trouble using a product (even with alternate access techniques) if the layout/organization of the information or product is too difficult to understand.<br><br>Many individuals will not be able to operate products, such as workstations, with sufficient efficiency to hold a competitive job if navigation is not efficient. | a) Make overall organization understandable (e.g., provide overview, table of contents, site maps, description of layout of device, etc.).<br>b) Don't mislead/confuse. (Be consistent in use of icons or metaphors. Don't ignore or misuse conventions.)<br>c) Allow users to jump over blocks of undesired information (e.g., repetitive information), especially if reading via sound or other serial presentation means.<br>d) Consider having different navigation models for novice vs expert users. |
| Facilitate understanding of content:<br>• without skill in the language used on the product (poor language skills or it is a second language for them)<br>• without good memory<br>• without background or experience with the topic | People with cognitive difficulties (or inexperienced users) may not be able to use devices or products with complex language. | a) Use the simplest, easiest to understand language and structure/format as is appropriate for the material/site/situation.<br>b) If phrases from a different language (than the rest of the page) are used in a document, either identify the language used (to allow translation) or provide a translation to the document language. |

**Table 18.1 (cont.)** Basic principles and strategies for access to electronic products and documents

| Basic Access Principle | Why | How—General |
|---|---|---|
| Provide compatibility with assistive technologies commonly used by people:<br>• with low vision<br>• without vision<br>• who are hard of hearing<br>• who are deaf<br>• without physical reach and manipulation<br>• who have cognitive or language disabilities | In many cases, a person coming up to a product will have assistive technology with them. If the person cannot use the product directly, it is important that the product be designed to allow them to use their assistive technology to access the product.<br><br>(This also applies to users of mobile devices, people with glasses, gloves or other extensions to themselves.) | a) Do not interfere with use of assistive technologies:<br>• personal aids (e.g., hearing aids)<br>• system-based technologies (e.g., OS features)<br><br>b) Support standard connection points for:<br>• audio amplification devices<br>• alternate input and output devices (or software)<br><br>c) Provide at least one mode where all functions of the product are controllable via ASCII/UNICODE input via an external port or via network connection. |

# References


Abascal, J., 1997, Ethical and social issues of the 'teleservices' for disabled and elderly people. In *An Ethical Global Information Society. Culture and Democracy Revisited*, edited by Berleur, J. and Whitehouse, D. (Chapman & Hall).

Abascal, J. and Nicolle, C., 2000, The use of USERfit methodology to teach usability guidelines. In *Tools for Working with Guidelines*, edited by Vanderdonckt, J. and Farenc, C. (London: Springer-Verlag), pp. 209–216.

Abernethy, C.N. and Lint, R., 1999, Technology standards to address those with disabilities. *Open Systems Standards Tracking Report, Compaq Computer Corporation Newsletter on Information Technology and Telecommunications Standardization*, **8**(5), May, pp. 1–4.

Akoumianakis, D. and Stephanidis, C., 1997a, Knowledge based support for user adapted interaction design. *Expert Systems with Applications*, **12**(1), pp. 225–245.

Akoumianakis, D. and Stephanidis, C., 1997b, Supporting user-adapted interface design: the USE-IT System. *International Journal of Interacting with Computers*, **9**(1), pp. 73–104.

Akoumianakis, D. and Stephanidis, C., 1999, Propagating experience-based accessibility guidelines to user interface development. *Ergonomics*, **42**(10), pp. 1283–1310.

Alpern, M., 1981, Color blind, colour vision. *Trends in Neurosciences*, **4**, pp. 131–134.

Anon., 1995, EJ497 680, Instructional design of computer-assisted instruction for use with students who have mild disabilities. *Teaching Exceptional Children*, **27**(3), pp. 77–79.

Apple Computer, 1989, Chapter one. [Videotape].

Audit Commission, 2000, *Fully Equipped—The Provision of Equipment to Older or Disabled People by NHS Trusts and Social Services Departments in England and Wales* (Audit Commission Publications).

Baåth, J., 1982, A list of ideas for the construction of distance education courses. In *Distance Education: A Short Handbook*, edited by Holmberg, B. (Malmo: Liber Hermods).

de Baenst-Vandenbroucke, A., Noirhomme-Fraiture, M., Lecomte, N., Patesson, R. and Steinberg, P., 1999, Design of a Web site with guidelines for accessibility. In *Workshop on Making Designers Aware of Existing Guidelines for Accessibility, INTERACT '99*, Edinburgh. To appear in *ACM SIGcaph Newsletter*. See also www.info.fundp.ac.be/IFIP13-3

Bailey, R.W., 1982, *Human Performance Engineering: A Guide for Systems Designers* (Englewood Cliffs, New Jersey: Prentice-Hall).

Bain, B.K., 1997, Evaluation of assistive technology devices and services. In *Advancement of Assistive Technology*, edited by Anogianakis, G., Buhler, C. and Soede, M. (Amsterdam: IOS Press), pp. 381–385.

Baine, D., 1982, *Instructional Design for Special Education* (Englewood Cliffs, New Jersey: Educational Technology Publications).

Bannon, L., 1991, From human factors to human actors: the role of psychology and human computer interaction studies in system design. In *Design at Work: Co-operative Design of Computer System*, edited by Greenbaum, J. and Kyng, M. (Hillsdale, NJ: Lawrence Erlbaum Associates), pp. 25–44.

Barnicle, K., 2000, On demand remote sign language interpretation. Poster session presented at the *World Wide Web 99 (WWW9) Conference*, Amsterdam.

Batavia, A.I. and Hammer, G.S., 1990, Toward the development of consumer-based criteria for the evaluation of assistive devices. *Journal of Rehabilitation Research and Development*, **27**(4), pp. 426–436.

Bergman, E. and Johnson E., 1995, Towards accessible human-computer interaction. In *Advances in HCI*, Vol.5, (Ablex).

Bjørneby, S., Collins, S., Nordby, K., Pereira, L.M. and Purificação, J., 1991, Attitudes and acceptance. In *Issues in Telecommunication and Disability, COST 219*, edited by Tetzchner, S. von (Luxembourg: Commission of the European Communities), pp. 110–119.

Bösser, T.J.H., Eggen, J.H., De Vet, J.H.M. and Westerink, J.H.D.M., 1995, Developments in information ergonomics. *IPO Annual Progress Report*, **30**, pp. 135–136.

Brandt, Å., 1996, *ICT Standardization and Disability in Europe*. Report of a European Policy Workshop, Amsterdam, edited by Brandt, Å. (Århus: Danish Centre for Technical Aids).

Brandt, Å., and Gjøderum, J., 1995, Older and disabled users participation in standardisation work is possible and valuable. In *The European Context for Assistive Technology. Proceedings of the 2nd TIDE Congress*, edited by Placencia-Porrero, I. and Puig de la Bellacasa, R. (Amsterdam: IOS Press), pp. 51–54.

Bryant, B.R. and Rivera, D.P., 1995, Using assistive technology to facilitate cooperative learning. Paper presented at *The 4th Conference of the Florida Assistive Technology Impact and the Technology and Media Division of the Council for Exceptional Children*, Orlando, Florida.

Buzan, T. and Buzan, B., 1993, *The Mind Map Book* (London: BBC Books).

Carruthers, S., Humphreys, A. and Sandhu, J.S., 1993, The market for R.T. in Europe: a demographic study of need. In *Rehabilitation Technology. Proceedings of the 1st TIDE Congress*, edited by Ballabio, E., Placencia-Porrero, I. and Puig de la Bellacasa, R. (Amsterdam: IOS Press), pp. 158–163.

CCTA, *Writing for the Web* (UK: CCTA), www.open.gov.uk/webguide/wgindex.htm

Cobut, G., Ekberg, J., Frederiksen, J., Leppo, A., Nordby, K. and Rollandi, G., 1991, Policy and legislation. In *Issues in Telecommunication and Disability, COST 219*, edited by Tetzchner, S. von, (Luxembourg: Office for Official Publications of the European Communities), pp. 71–81.

Colwell, C. and Petrie, H., 1999, A preliminary evaluation of the WAI guidelines for producing accessible Web pages. In *Assistive Technology on the Threshold of the New Millennium*, edited by Bühler, C. and Knops, H. (Amsterdam: IOS Press).

*Common Desktop Environment: Style Guide and Certification Checklist* (Hewlett-Packard, IBM, Novell and Sun Microsystems).

Conrad, J.G., 1994, *A System for Discovering Relationships by Feature Extraction from Text Databases.* Center for Intelligent Information Retrieval (CIIR) (Amherst, USA: University of Massachusetts).

Council of the European Communities, 1990, *Directive 90/270/EEC on the Minimum Safety and Health Requirements for Work with Display Screen Equipment* (Brussels: CEC).

Cress, C.J. and Goltz, C.C., 1989, Cognitive factors affecting accessibility of computers and electronic devices. In *RESNA 12th Annual Conference*, New Orleans, Lousiana, Section 3.6, pp. 25–26.

Disabilities Services, 1994, *Competency Assessment Mechanisms and Reasonable Adjustment: A Report to the Department of Employment, Education and Training* (New South Wales, Australia: NSW TAFE Commission).

Docx, S., Evenepoel, F. and Engelen, J., 1998, *Access to Multimedia and Hypertext by Print Disabled People,* www.esat.kuleuven.ac.be/teo/docarch/toegankelijkheid/rapport%20toegankelijkheid.doc

Downie, A., 1993a, *Adaptive Technology Survey: Equipment for People who are Deaf or Hearing Impaired* (Redfern, NSW: OTEN Occasional Papers 7).

Downie, A., 1993b, *Adaptive Technology Survey: Equipment for People who are Blind or Vision Impaired* (Redfern, NSW: OTEN Occasional Papers 8).

Downie, A., 1995, *Adaptive Technology Survey: Equipment for People who have a Physical Disability* (Redfern, NSW: OTEN Occasional Papers 9).

Downie, A., 1996, *Adaptive Technology Survey: Equipment for People who have Intellectual, Neurological or Learning Disabilities* (Redfern, NSW: OTEN Occasional Papers 10).

Ekberg, J., Lindström, B. and Nordby, K., 1991, Standardisation. In *Issues in Telecommunication and Disability, COST 219*, edited by Tetzchner, S. von (Luxembourg: Office for Official Publications of the European Communities), pp. 174–178.

European Commission, 1996, White Paper—Equal Opportunities and Non-Discrimination for People with Disabilities, Council Decision of 20 December 1996. *Official Journal* **CO12**, 13 Jan. 1997.

EUROSTAT 1992, *Rapid Reports, Population and Social Conditions: Disabled People Statistics* (Luxembourg: Commission of the European Communities).

Foreman, J., 1991, San Francisco passes ordinance regulating VDT use. *Archeological Ophthalmology*, **109**, pp. 477.

Foster, G.T. and Solberg, L.A., 1997, The ARIADNE Project. Access, navigation and information services in the labyrinth of large buildings. In *Advancement of Assistive Technology*, edited by Anogianakis, G., Buhler, C. and Soede, M. (Amsterdam: IOS Press), pp. 211–216.

Frederiksen, J. *et al*, 1989, *Use of Telecommunications: The needs of People with Disabilities* (EEC DG XIII/097/89.EUCO TELE219/NPH/89), Diskette version March 1990, Audio version March 1990, Spanish version June 1989, French version June 1991, Dutch version September 1991, German version November 1991.

Fuhr, N. and Buckley, C., 1991, A probabilistic learning approach for document indexing. *ACM Transactions on Information Systems*, **9**(3), pp. 223–248.

Furugren, B. and Lundman, M., 1995, New roles for disabled users in technology application work. In *The European Context for Assistive Technology, Proceedings of the 2nd TIDE Congress*, edited by Placencia-Porrero, I. and Puig de la Bellacasa, R. (Amsterdam: IOS Press), pp. 33–37.

Furuta, R., Plaisant, C. and Shneiderman B., 1989, Automatically transforming regularly structured linear documents into Hypertext. *Electronic Publishing*, **2**(4), pp. 211–229.

Gann, D., Barlow, J. and Venables, T., 1999, *Digital Futures: Making Homes Smarter* (Chartered Institute of Housing).

Gilbert, J.G., 1967, Age changes in colour matching. *Journal of Gerontology*, **12**, pp. 210–215.

Gill, J., 1997, *Access Prohibited? Information for Designers of Public Access Terminals*, www.tiresias.org/pats

Gill, J. and Shipley, T., 1997, *Disaster or Opportunity? The Impact of Telecommunications Deregulation on People with Disabilities* (London: Royal National Institute for the Blind).

Gjøderum, J. (editor), 1995, Telecommunications and people with disabilities: legislation and standardisation. In *Proceedings of the COST 219 Seminar*, Budapest (Esbjerg, Denmark: Rosendahls).

Goffee, N., 1996, *An Evaluation of the USERfit Handbook in Relation to Service Design in the Assistive Technology Sector*. Unpublished MSc IT thesis, Loughborough University.

Goffinet, L. and Noirhomme, M., 1996, *Automatic Hypertext Link Generation based on Similarity Measures between Documents* (Institut d'Informatique, FUNDP, RP–96–034), www.fundp.ac.be/~lgoffine

Goldsmith, S., 1984, *Designing for the Disabled* (RIBA Publications).

Goor, M.B. and Schwenn, J.O., 1993, Accommodating diversity and disability with cooperative learning. *Intervention in School and Clinic*, **29**(1), pp. 6–16.

Hector , C., 2000, *RTFtoHTML Software*, www.sunpack.com/RTF/

Hofmeister, A.M. *et al*, 1992, Learner diversity and instructional video: implications for developers. *Educational Technology*, **32**(7).

Hollins, M., 1989, *Understanding Blindness: An Integrative Approach*, (Hillsdale, NJ: Lawrence Erlbaum).

Hooyman, N.R. and Kiyak, H.A., 1988, *Social Gerontology: A Multidisciplinary Perspective*, (Boston: Allyn and Bacon Inc.)

International Organisation for Standardisation, 1998, *ISO 9241 Software Ergonomics with Visual Display Terminals (VDTs), Part 11 Guidance on Usability. International Standard* (Switzerland: International Organisation for Standardisation).

International Organisation for Standardisation, 1999, *ISO 13407 Human Centred Design Process for Interactive Systems. International Standard* (Switzerland: International Organisation for Standardisation).

International Organisation for Standardisation, 2000, *ISO 16071 Ergonomics of Human–System Interaction—Guidance on Software Accessibility. Technical specification* (Switzerland: International Organisation for Standardisation).

Isaacs, E.A., Tang, J.C., Foley, J., Johnson, J., Kuchinsky, A., Scholtz, J. and Bennett, J., 1996, Technology transfer: so much research, so few good products. In *Companion Proceedings of CHI '96* (New York: ACM SIGCHI), pp. 155–156.

Jain, A.K. and Usiak, D., 1997, Customer orientation: a blueprint for action. In *Proceedings of the RESNA '97 Annual Conference, Let's Tango—Partnering People and Technology*, Pittsburgh, Pennsylvania, (RESNA Press), pp. 136–138.

Janssen, H.T.J. and van der Vegt, H., 1998, Commercial design for all. In *Improving the Quality of Life for the European Citizen. Proceedings of the 3rd TIDE Congress*, edited by Placencia-Porrero, I. and Ballabio, E. (Amsterdam: IOS Press), pp. 84–87.

Kemppainen, E., Gjøderum, J. and Martin, M., 1995, Legislation as a support for telecommunications availability and access. In *Telecommunications for all, COST 219*, edited by Roe, P.R.W. (Luxembourg: Office for Official Publications of the European Communities), pp. 81–89.

Koucelik, J.A., 1982, *Aging and the Product Environment* (New York: Van Nostrand Reinhold).

Kouroupetroglou, G., Viglas, C., Anagnostopoulos, A., Stamatis, C. and Pentaris, F., 1996, A novel software architecture for computer-based interpersonal communication aids. In *Interdisciplinary Aspects on Computers Helping People with Special Needs. Proceedings of the 5th International Conference on Computers and Handicapped People (ICCHP '96)*, Linz, Austria, edited by Klaus, J., Auff, E., Kremser, W. and Zagler, W. (Vienna, Austria: Oldenbourg), pp. 715–720.

Kumar, S. (editor), 1997, *Perspectives in Rehabilitation Ergonomics*, (London: Taylor & Francis).

Kwok, K.L., 1995, A network approach to probabilistic information retrieval. *ACM Transactions on Information Systems*, **13**(3).

Landau, F.H., Hanley, M.N. and Hein, C.M., 1998, Application of existing human factors guidelines to ATIS (advanced traveler information systems). In *Human Factors in Intelligent Transportation Systems*, edited by Barfield, W. and Dingus, T.A. (Mahwah, New Jersey: Lawrence Erlbaum Associates), pp. 397–443.

Legge, G.E., Rubin, G.S., Pelli, D.G. and Schleske, M.M., 1985, Psychophysics of reading—II. Low vision. *Vision Research*, **25**(2), pp. 253–266.

Lehnert, W.G., 1992, Automating the construction of a Hypertext system for scientific literature. In *Proceedings of the AAAI Workshop on Communicating Scientific and Technical Knowledge*.

Lindström, J.-I. (editor), 1998, *Universal Services Issues* (Brussels).

*Macintosh Human Interface Guidelines*, 1992 (Addison-Wesley).

MacLean, A., Young, R.M., Bellotti, V. and Moran, T., 1991, Questions, options, and criteria: elements of design space analysis. *Human-Computer Interaction*, **6**(3&4), pp. 201–250.

MacMillan, D.L., Keogh, B.K. and Jones, R.L., 1986, Special educational research on mildly handicapped learners. In *Handbook of Research on Teaching,* 3rd ed., edited by Wittrock, M.C. (New York: Macmillan).

Mann, W.C., Ottenbacher, K.J., Tomita, M.R. and Packard, M.S., 1994, Design of hand held remotes for older persons with impairments. *Assistive Technology*, **6**, pp. 140–146.

Marin-Lamellet, C.M., Burnett, G. and Nicolle, C., 1999, *A Methodology for the Assessment of Telematic Applications for Travellers who are Elderly or Disabled* (CEC Transport Telematics Project TR 1108, TELSCAN Deliverable 4.2).

Martin, M., 1991, People with special needs as a market. In *Issues in Telecommunication and Disability, COST 219*, edited by Tetzchner, S. von (Luxembourg: Office for Official Publications of the European Communities), pp. 55–60.

Martin, M., Gjøderum, J. and Kemppainen, E., 1995, The principles and practice of standardization. In *Telecommunications for all, COST 219*, edited by Roe, P.R.W. (Luxembourg: Office for Official Publications of the European Communities), pp. 63–80.

McCann, A., 1997, Designing accessible learning materials for learners with disabilities and learning difficulties. In *Workshop on Guidelines for the Design of HCI for People with Disabilities, INTERACT'97*, Sydney.

Microsoft Corporation, Accessibility and Disabilities Group, 1995, *Designing Accessible Applications* (Redmond: Microsoft).

Miller, G.A., 1956, The magical number seven, plus or minus two: some limits on our capacity for processing information. *Psychological Review*, **63**, pp. 81–97.

Morgan, J.W., 1993, Assessment and training of people with disabilities using new technologies. Human welfare and technology. In *Proceedings of the Husita 3 Conference on IT and the Quality of Life and Services*, Maastricht, The Netherlands (Van Gorcum), pp. 133–137.

Morris, J.M., 1994, User interface design for older adults. *Interacting with Computers*, **6**(4), pp. 373–393.

Nes, F.L. van, 1994, Developments in information ergonomics. *IPO Annual Progress Report*, **29**, pp. 137–140.

Nes, F.L. van and Bouma, H., 1990, Telecommunication interfaces for the visually disabled. In *Proceedings of the 13th International Symposium on Human Factors in Telecommunications, HFT '90,* Torino, Italy, pp. 227–234.

Neugarten, B., 1974, Age groups in American society and the rise of the young-old. *The Annals of the American Academy of Political and Social Science. Political Consequences of Aging*, **415**, pp. 187–198.

Nicolle, C. and Burnett, G. (editors), 1999, *TELSCAN Code of Good Practice and Handbook of Design Guidelines for Usability of Systems by Elderly and Disabled Travellers* (CEC Transport Telematics Project TR 1108, TELSCAN Deliverable 5.2), http://hermes.civil.auth.gr/telscan/telsc.html

Nicolle, C., Burnett, G., Ross, T., Ståhl, A., Petzäll, J., Veenbaas, R., Hekstra, A., Marin-Lamellet, C., Oxley, P., Barham, P., Simões, A., Naniopoulos, A., 1998, *Inventory of ATT System Requirements of Elderly and Disabled Drivers and*

*Travellers* (CEC Transport Telematics Project TR 1108, TELSCAN Deliverable 3.1), http://hermes.civil.auth.gr/telscan/telsc.html

Nicolle, C. and Peters, B., 1999, Elderly and disabled travellers: ITS designed for the 3rd Millennium. *Transportation Human Factors*, **1**(2), pp. 121–134.

Nicolle, C. and Richardson, S.J., 1995, Defining user requirements for people with dementia who wander. In *Proceedings of ECART 3* (Lisbon: National Secretariat of Rehabilitation).

Nicolle, C., Veenbaas, R. and Ross, T., 1997, Using the travelling task as a tool to define ITS requirements for elderly and disabled people. In *Proceedings of the 4th World Congress on Intelligent Transport Systems* (CD ROM), (Berlin: Ertico).

Nielsen, J., 1993, *Usability Engineering* (Morgan Kaufmann).

Nordby, K., 1991, Proposals for standardisation activities. In *Issues in Telecommunication and Disability, COST 219*, edited by Tetzchner, S. von, (Luxembourg: Office for Official Publications of the European Communities), pp. 179–194.

Nordic Cooperation on Disability, 1998, *Nordic Guidelines for Computer Accessibility*, 2nd ed., edited by Thorén, C.

Okolo, C.M., Bahr, C. and Reith, H., 1993, A retrospective view of computer-based instruction. *Journal of Special Education Technology* **12**(1), pp. 1–27.

Pascoe, J., Pain, H., McLellan, D.L., Jackson, S. and Ballinger, C., 1995, Involving users in the evaluation of assistive equipment. In *The European Context for Assistive Technology. Proceedings of the 2nd TIDE Congress*, edited by Placencia-Porrero, I. and Puig de la Bellacasa, R. (Amsterdam: IOS Press), pp. 114–117.

Perret, B., 1995, Marketing considerations: the BT approach. In *Telecommunications for all, COST 219*, edited by Roe, P.R.W. (Luxembourg: Office for Official Publications of the European Communities), pp. 41–46.

Petrie, H., Morley, S., McNally, P., O'Neill, A.-M. and Majoe, D., 1997, Initial design and evaluation of an interface to hypermedia systems for blind users. In *Proceedings of Hypertext '97*, Southampton, UK (New York: ACM Press), pp. 48–56.

Pheasant, S., 1986, *Bodyspace: Anthropometry, Ergonomics and Design* (London: Taylor & Francis).

Poll, L.H.D., 1996, *Visualising Graphical User Interfaces for Blind Users*, Doctoral dissertation (Eindhoven: Eindhoven University of Technology Press).

Poulson, D.F., 1997, Using new technology to support the provision of care services. *Computing & Control Engineering Journal*, **8**(5), Oct., pp. 203–207.

Poulson, D.F. and Richardson, S.J., 1994, Developing adaptable smarter homes for elderly and visually impaired people. In *Ergonomics and Design. IEA '94. Proceedings of the 12th Triennial Congress of the International Ergonomics Association*, Vol. 4, Toronto, Canada, pp. 56–58.

Poulson, D.F. and Richardson, S., 1998, USER*fit*—a framework for user-centred design in assistive technology. *Technology and Disability,* **9**(198), pp. 163–171.

Poulson, D., Ashby, M. and Richardson, S. (editors), 1996, *USERfit—A Practical Handbook on User-Centred Design for Assistive Technology* (Brussels, Luxemburg: ECSC–EC–EAEC).

Preece, J., Rogers, Y., Sharp, H., Benyon, D. and Carey, T., 1994, *Human–Computer Interaction* (Wokingham, England: Addison–Wesley), pp. 487–500.

Quirk, R., 1994, *Language and Literacy Issues in Competency Based Assessment.* Prepared for the Assessment Centre for Vocational Education; commissioned by the Foundation Studies Training Division, TAFE NSW, utilising funding supplied by the Education and Training Foundation, Australia.

Rahman, M. and Sprigle, S., 1997, Physical accessibility guidelines of consumer product controls. *The Assistive Technology Journal*, **9**, pp. 3–14.

Raschko, B.B., 1991, *Housing Interiors for the Disabled and Elderly* (New York: Van Nostrand Reinhold).

Rijsbergen, C.J. van, 1979, *Information Retrieval* (London: Butterworths).

Riley, M.W. and Riley, J., 1986, Longevity and social structure: the potential of the added years. In *Our Aging Society: Paradox and Promise*, edited by Pifer, A. and Bronte, L. (New York: W.W. Norton).

Ripps, H., 1982, Night blindness revisited: from man to molecules. *Investigative Ophthalmology and Visual Science*, **23**, pp. 588–609.

Roe P. (editor), 1995, *Telecommunications for All* (Brussels/Luxembourg: ECSC–EC–EAEC, CD–90–95–712–EN–C), Spanish version 1996 (Fundesco).

Romich, B., 1993, Assistive technology and AAC: an industry perspective. *Assistive Technology* **5**, pp. 74–77.

Salminen, A., Tague–Sutcliffe, J. and McCellan, C., 1995, From text to Hypertext by indexing. *ACM Transactions on Information Systems*, **13**(1).

Salton, G., 1989, *Automatic Text Processing* (Reading, Massachusetts: Addison-Wesley).

Sanders, M.S. and McCormick, E.J., 1992, *Human Factors in Engineering and Design*, 6th ed. (New York: McGraw Hill).

Sandhu, J., 1993, Design for the elderly: user-based evaluation studies involving elderly users with special needs. *Applied Ergonomics* **24**(1), pp. 30–34.

Sandhu, J.S., 1998, What is design for all? In *Improving the Quality of Life for the European Citizen. Proceedings of the 3rd TIDE Congress*, edited by Placencia-Porrero, I. and Ballabio, E. (Amsterdam: IOS Press), pp. 88–91.

Sarre, F. and Güntzer, U., 1991, Automatic transformation of linear text into Hypertext. In *Proceedings of DASFAA '91*, Tokyo, pp. 498–506.

Savidis, A., Stephanidis, C. and Akoumianakis, D., 1997, Unifying toolkit programming layers: a multi-purpose toolkit integration module. In *Proceedings of the 4th Eurographics Workshop on Design, Specification and Verification of Interactive Systems. DSV–IS '97*, Granada, Spain, edited by Harrison, M.D. and Torres, J.C. (Berlin: Springer-Verlag), pp. 177–192.

Savidis, A., Stergiou, A. and Stephanidis, C., 1997, Generic containers for metaphor fusion in non-visual interaction: The HAWK Interface Toolkit. In *Proceedings of the 6th International Conference on Man–Machine Interaction Intelligent Systems in Business. Interfaces '97*, Montpellier, France, edited by Rault, J.–C. *(La Lettre de l'IA: The Advanced Information Technology Newsletter*), pp. 194–196.

Savidis, A., Vernardos, G. and Stephanidis, C., 1997, Embedding scanning techniques accessible to motor-impaired users in the Windows Object Library. In *Design of Computing Systems: Cognitive Considerations. Proceedings of the 7th*

*International Conference on Human–Computer Interaction. HCI International '97*, Vol. 1, San Francisco, USA, edited by Salvendy, G., Smith, M.J. and Koubek, R.J. (Amsterdam: Elsevier Science), pp. 429–432.

Scadden, L.A. and Vanderheiden, G.C., 1988, *Considerations in the Design of Computers and Operating Systems to Increase their Accessibility to Persons with Disabilities. Design Considerations Task Force. Computers and Operating Systems* (Trace Center).

Schaie, K.W., 1983, *Longitudinal Studies of Adult Psychological Development and Adult Development and Aging* (New York: Guilford Press).

Small, A., 1987, Design for older people. In *Handbook of Human Factors*, edited by Salvendy, G. (New York: Wiley).

Smith, J.W., 1996, *ISO and ANSI Ergonomic Standards for Computer Products. A Guide to Implementation and Compliance* (New Jersey: Prentice-Hall).

Smith, S. and Mosier, J., 1986, *Guidelines for Designing User Interface Software* (Bedford, Massachusetts: MITRE Corporation, ESD–TR–86–278).

Sony, 1998, *Sony barrier-free charter, Sony Product development guidelines*, www.rnib.org.uk/wesupply/products/sony.htm

Spiezle, C.D., 1999, *Effective Web Design Considerations for Older Adults*, (Redmond: Microsoft Corporation).

Stephanidis, C. (editor), 1994, *The Guy Cobut report on 'Universal Access to Telecommunications services in Europe'* (Brussels).

Stephanidis, C., 1997, Designing for accessibility: reflections on the use of human factors design guidelines. In *Workshop on Guidelines for the Design of HCI for People with Disabilities, INTERACT '97*, Sydney.

Stephanidis, C., Salvendy, G., Akoumianakis, D., Arnold, A., Bevan, N., Dardailler, D., Emiliani, P.L., Iakovidis, I., Jenkins, P., Karshmer, A., Korn, P., Marcus, A., Murphy, H., Oppermann, C., Stary, C., Tamura, H., Tscheligi, M., Ueda, H., Weber, G. and Ziegler, J., 1999, Toward an information society for all: HCI challenges and R&D recommendations. *International Journal of Human–Computer Interaction*, **11**(1), pp. 1–28.

Sterling, L., 1994, *Students with Acquired Brain Injuries in Primary and Secondary Schools.* Draft report of the project by the Head Injury Council of Australia funded by the Commonwealth Department of Employment, Education and Training, Canberra, Australia.

Stopford, V., 1987, *Understanding Disability: Causes, Characteristics and Coping* (London: Edward Arnold).

Story, M.F., 1998, Maximising usability: the principles of universal design. *The Assistive Technology Journal*, **10**(1), pp. 4–12.

Stowe, J., Rowley, C. and Chamberlain, M.A., 1988, Acquisition and use of communication aids by those buying directly from the supplier. *British Journal of Occupational Therapy*. **51**(3), Mar., pp. 97–100.

Tahkokallio, P., 1998, Through other eyes. From knowledge to understanding. In *Improving the Quality of Life for the European Citizen. Proceedings of the 3rd TIDE Congress*, edited by Placencia-Porrero, I. and Ballabio, E. (Amsterdam: IOS Press), pp. 63–66.

Tetzchner, S. von, and Nordby, K., 1991, Telecommunication behaviour. In *Issues in Telecommunication and Disability, COST 219*, edited by Tetzchner, S. von (Luxembourg: Commission of the European Communities), pp. 26–38.

Trace Center, 1997, *Unified Web Site Accessibility Guideline, Advanced Draft Copy. Page Author, User Agent and Assistive Technology Guidelines and Issues—Version 8* (Madison, WI: Trace Center), http://trace.wisc.edu/

Tran, A., Caza, G., Gross, E. and Ramarao, S., 1997, Design of the Bowling Empowerment Tool. In *Proceedings of the RESNA '97 Annual Conference, Let's Tango—Partnering People and Technology*, Pittsburgh, Pennsylvania (RESNA Press), pp. 497–499.

Ueno, K. and Ogawa, K., 1993, A design guideline search method that uses a neural network. In *Proceedings of HCI'93*.

Ulrich, K.T. and Eppinger, S.D., 1995, *Product Design and Development* (New York: McGraw-Hill).

Vanderdonckt, J., 1995, Accessing guidelines information with SIERRA. In *Proceedings of Interact '95* (Chapman & Hall), pp. 311–316.

Vanderdonckt, J. and Farenc, C. (eds.), 2000, *Tools for Working with Guidelines*, (London: Springer-Verlag).

Vanderheiden, G.C., 1992a, *Application Software Design Guidelines: Increasing the Accessibility of Application Software to People with Disabilities and Older Users, Version 1.1* (Madison, Wisconsin: Trace Research and Development Center).

Vanderheiden, G.C., 1992b, *Making Software More Accessible for People with Disabilities: Release 1.2* (Madison, Wisconsin: Trace Research and Development Center).

Vanderheiden, G.C., 1992c, A brief look at technology and mental retardation in the 21st century. In *Mental Retardation in the Year 2000*, edited by Rowitz, L. (New York: Springer-Verlag), pp. 268–278.

Vanderheiden, G.C., 1995, Access to global information infrastructure (GII) and next-generation information systems. In *Proceedings of the 18th International Congress on Education of the Deaf*, Tel Aviv, Israel, edited by Weisel, A. (Tel Aviv: Ramot Publications Tel Aviv University).

Vanderheiden, G.C. and Lee, C.C. (Coordinators), 1988, *Considerations in the Design of Computers to Increase Their Accessibility by Persons with Disabilities* (Madison, WI: Trace Center, Industry/Government Cooperative Initiative on Computer Accessibility Task Force).

Vanderheiden, G.C., Chisholm, W.A., Ewers, N. and Dunphy, S., 1997, *Unified Web Site Accessibility Guidelines, Version 7.2, June 1997* (Trace R and D Center, University of Wisconsin–Madison).

Veenbaas, R., 2000, Travel information checklist. In Börjesson, M., Nicolle, C. and Veenbaas, R., *Updating of User Requirements of Elderly and Disabled Drivers and Travellers* (CEC Transport Telematics Project TR 1108, TELSCAN Deliverable 3.3).

Velasco, C.A., 1998, The Information Society disAbilities Challenge (ISdAC): paving the way for the active enduser. In *Improving the Quality of Life for the*

*European Citizen. Proceedings of the 3rd TIDE Congress*, edited by Placencia-Porrero, I. and Ballabio, E. (Amsterdam: IOS Press), pp. 473–477.

Vet, J.H.M. de, 1993, User–interface specification guidelines for consumer electronics products. *IPO Annual Progress Report*, **28**, pp. 151–159.

Vet, J.H.M. de, 1996, Developments in user-centred design. *IPO Annual Progress Report*, **31**, pp. 19–20.

Welford, A.T., 1985, Changes of performance with age: an overview. In *Aging and Human Performance*, edited by Charness, N. (New York: Wiley).

Whitlock, A., 1998, *The Evaluation of the Use of USERfit, a Handbook on User-Centred Design for Assistive Technology*. Unpublished final year project in Ergonomics, Department of Human Sciences, Loughborough University.

Wieck, J., Perez, J. and Meyyappan, R., 1997, Design and development of a self-orienting 'smart' spoon. In *Proceedings of the RESNA '97 Annual Conference, Let's Tango—Partnering People and Technology*, Pittsburgh, Pennsylvania (RESNA Press), pp. 488–490.

World Health Organization, 1980, *International Classification of Impairments, Disabilities, and Handicaps* (Geneva: World Health Organization).

World Health Organization, 1999, *International Classification of Functioning and Disability (Beta-2 Draft)* (Geneva: World Health Organization).

World Wide Web Consortium (W3C), 1999, *Web Content Accessibility Guidelines 1.0*, edited by Chisholm, W., Vanderheiden, G., and Jacobs, I., www.w3.org/TR/WAI-WEBCONTENT/

Wünschmann, W., 2000, Structure and content of accessibility guidelines. In *Proceedings of the International Conference on Computers Helping People with Special Needs. ICCHP2000*, Karlsruhe (Austrian Computer Society (OCG)).

# World Wide Web Sites

All links were checked during the summer of 2000 and they were available at that time.

Adobe Acrobat Reader
http://access.adobe.com

Alliance for Technology Access
www.ataccess.org/design.html

Americans with Disabilities Act
www.usdoj.gov/crt/ada/adahom1.htm

Apple Macintosh and people with special needs
www.apple.com/education/k12/disability/

AVANTI project and browser
www.avanti-acts.org/

Bobby
www.cast.org/bobby/

COST219 and COST 219bis
www.stakes.fi/cost219
www.cost219.org

*Disability Discrimination Act*, 1995, issued on behalf of the Minister for Disabled People
www.disability.gov.uk/dda/index.html

eEurope
http://europa.eu.int/comm/information_society/eeurope/actionplan/actline2c_en.htm
http://europa.eu.int/comm/information_society/eeurope/index_en.htm
http://europa.eu.int/comm/information_society/eeurope/pdf/actionplan_en.pdf
www.egroups.com/message/eeurope-pwd/36?

European Union's *Addressing the Digital Divide*
europa.eu.int/rapid/start/cgi/guesten.ksh?p_action.gettxt=gt&doc=IP/00/477|0|RAPID&lg=EN

FORTUNE project
www.fortune-net.org

HARMONY project
www.esat.kuleuven.ac.be/teo/harmony

HELPDB (Support for standard browsers)
www.dinf.org/csun_99/session0087.html

Hypertext guidelines
www.info.fundp.ac.be/httpdocs/guidelines/

INCLUDE project
www.stakes.fi/include

International Federation for Information Processing (IFIP)
www.ifip.or.at/

IFIP Working Group 13.3 on Human–Computer Interaction and Disability
www.info.fundp.ac.be/IFIP13-3/

International Organisation for Standardisation (ISO)
www.iso.ch/infoe/intro.htm

Lynx browser
http://lynx.browser.org/

Microsoft Windows® Guidelines for Accessible Software Design
www.microsoft.com/enable/dev/guidelines/software.htm

Mind maps©
http://ourworld.compuserve.com/homepages/marco_cosimetti/link2.htm

National Center for Accessible Media–Web Accessibility Symbol
www.psc-cfp.gc.ca/dmd/access/welcome1.htm

Public access terminals
www.tiresias.org/pats

Royal National Institute for the Blind UK (RNIB)
www.rnib.org.uk/digital

Screen fonts
www.tiresias.org/fonts

Screenphones
www.rnib.org.uk/wedo/research/sru/phones.htm

Sony, 1998, *Sony barrier-free charter, Sony Product development guidelines*, www.rnib.org.uk/wesupply/products/sony.htm

Standard Rules on the Equalization of Opportunities for Persons with Disabilities
www.dinf.org/un_dinf/un_4896.htm

Telephones
www.tiresias.org/phoneability/telephones

TELSCAN project—systems for older and disabled travellers
http://hermes.civil.auth.gr/telscan/telsc.html

Text Telephones (TTYs)
www.ultratec.com

TIDE WAI project
www2.echo.lu/telematics/disabl/wai.html

Trace Research and Development Center
trace.wisc.edu
http://tracecenter.org/docs/consumer_product_guidelines/consumer.htm
*Web Access Guidelines*. No longer available at www.trace.wisc.edu, replaced by W3C–WAI guidelines

Universal Design
trace.wisc.edu/docs/whats_ud/whats_ud.htm

US Access Board
www.access-board.gov/

US Department of Education World Wide Web (WWW) Server Standards and Guidelines
http://inet.ed.gov/~kstubbs/wwwstds.html

Web Page Accessibility Self-Evaluation Test
www.psc-cfp.gc.ca/dmd/access/welcome1.htm

*World Wide Web Accessibility to People with Disabilities; A Usability Perspective*
www.staff.uiuc.edu/~jongund/access-overview.html

World Wide Web Consortium–Web Accessibility Initiative (W3C–WAI)
www.w3.org/WAI/References
www.w3.org/WAI/References/Browsing

*Authoring Tool Accessibility Guidelines*
www.w3.org/TR/ATAG

The Authoring Tool Accessibility Guidelines Checklist
www.w3.org/TR/2000/REC-ATAG10-20000203/atag10-chklist.html

SMIL specification
www.w3.org/TR/REC-smil/

*Statement before US House of Representatives*
www.house.gov/judiciary/brew0209.htm

*User Agent Accessibility Guidelines*
www.w3.org/TR/UAAG

The User Agent Guidelines Checklist
www.w3.org/TR/WAI-USERAGENT/full-checklist.html

*Web Content Accessibility Guidelines*
www.w3.org/TR/WCAG

The Web Content Guidelines Checklist
www.w3.org/TR/WAI-WEBCONTENT/full-checklist.html

# Index

Tables and figures are indicated by t and f as appropriate after page references